Clarke Willmott & Clarke

CONTAMINATED LAND

CONTAMINATED LAND

Tom Graham, BA (Hons)
Solicitor

JORDANS

1995

Published by
Jordan Publishing Limited
21 St Thomas Street
Bristol BS1 6JS

British Library Cataloguing-in-Publication Data

A catalogue record for this book is available from the British Library.

ISBN 0 85308 267 7

Photoset by Pentacor plc
Printed in Great Britain by Bookcraft (Bath) Ltd, Midsomer Norton

ACKNOWLEDGEMENTS

I once read somewhere (in a somewhat antique volume of one of Cheshire's property law books, I think) that the author's ideal is to write his manuscript, then leave it for some time and come back to it with a fresh mind. I agree in principle but, alas, modern environmental law moves at such a pace that he is not allowed to apply that salutary advice – he is trying to hit a very rapidly moving target. Instead, one relies on the fresh minds of one's friends and colleagues, who, as usual, will always manage to pick up slips of the pen and obscure reasoning. It would take too long to mention specific names, because I have enjoyed considerable help from my friends whilst writing this book, so, save for two exceptions, I thank them collectively. The first exception is my beloved wife, Linda, who has had to put up with my vanishing act into the study all too frequently over the past two years – much love and many thanks to her. The second exception is my colleague, Rachael Keary, for her help in finalising the manuscript and putting up with my awful jokes.

The publishers and I would like to thank the following: the Controller of HMSO for permission to reproduce the material on pp 147, 150 and 151, which is Crown copyright; the National Rivers Authority for permission to reproduce the material on pp 219–220; and the Welsh Development Agency for permission to reproduce the material on pp 224–227. I would also like to acknowledge the debt I owe to other writers in completing this work. I have found the ENDS Report (published by Environmental Data Services Ltd) to be a particularly fecund source of timely information; an article by Alan Watson was a great help in Chapter 2; likewise an article by William Birtles in relation to Chapter 15.

Of course, as always, the writer has only himself to blame for any gaffes. I hope they are few and far between and that this modest tome proves to be both readable and instructive.

PREFACE

Environmental law has spawned some extremely encyclopaedic written works in very recent years and it would have been extremely courageous of the writer to attempt to duplicate these feats. Instead, I have focused on an area which is of particular concern to me, an area which I know is also of great concern to those involved in land: namely, contaminated land. This book gives, I hope, practical guidance on the ownership of, and dealings in, contaminated land with particular emphasis on its development.

The Environment Act 1995 came onto the statute book at the same time as the final proof of this book was being settled and so Chapter 16 was added as a self-contained summary of its provisions. I suspect that this book is likely to be the first non loose-leaf book on the market dealing with the new Act and so I hope Chapter 16 will provide early guidance as to its operation in respect of contaminated land.

The law is stated as at 1 August 1995.

T. GRAHAM
Bristol
September 1995

CONTENTS

TABLE OF CASES

References are to paragraph numbers.

TABLE OF STATUTES

References are to paragraph numbers.

TABLE OF STATUTORY INSTRUMENTS

References are to paragraph numbers.

TABLE OF EC LEGISLATION

References are to paragraph numbers.

ABBREVIATIONS

AEC	Association of Environmental Consultancies
ASTM	American Society for Testing Materials
BCN	Breach of condition notice
BNF	British Nuclear Fuels
BSI	British Standards Institute
CCME	Canadian Council of Ministers for the Environment
CLR No 1	DoE Contaminated Land Research Report, *A Framework for Assessing the Impact of Contaminated Land on Groundwater and Surface Water* (2 vols)
CLR No 2	DoE Contaminated Land Research Report, *Guidance on Preliminary Site Inspection of Contaminated Land* (2 vols)
CLR No 3	DoE Contaminated Land Research Report, *Documentary Research on Industrial Sites*
CLR No 4	DoE Contaminated Land Research Report, *Sampling Strategies for Contaminated Land*
COPA	Control of Pollution Act 1974
COREPER	Committee of Permanent Representatives
CWC	Cambridge Water Company
DoE	Department of the Environment
EARA	Environmental Auditors Registration Association
ECL	Eastern Counties Leather plc
ECSC	European Coal and Steel Community
EPA	Environmental Protection Act 1990
Euratom	European Atomic Energy Community
GDO	Town and Country Planning (General Permitted Development) 1995 Order
HMIP	Her Majesty's Inspectorate of Pollution
HSA	Hazardous Substances Authority
HSC	Hazardous substances consent
ICRCL	Inter-departmental Committee on the Redevelopment of Contaminated Land
LCA	Local Commissioner for Administration
LGC	Laboratory of the Government Chemist
LNR	Local nature reserve
LPA	Local planning authority
MPG	Mineral Planning Guidance Note
NCO	Nature conservation order
NHBC	National House Building Council
NNR	National nature reserve
NPACA	National Parks and Access to Countryside Act 1949

NRA	National Rivers Authority
P(HS)A	Planning (Hazardous Substances) Act 1990
PCA	Parliamentary Commissioner for Administration
PCE	Perchloroethene
PPG	Planning Policy Guidance Note
PPG9	Planning Policy Guidance Note 9, *Nature Conservation* (DoE, 1994)
QMV	Qualified Majority Voting
RPG	Regional Planning Guidance Note
SAC	Special area of conservation
SEA	Single European Act 1986
SPA	Special protection area
SSSI	Site of special scientific interest
TCPA	Town and Country Planning Act 1990
TPO	Tree preservation order
UCO	Town and Country Planning (Use Classes) Order 1987
UDP	Unitary development plan
WCA	Wildlife and Countryside Act 1981
WDA	Welsh Development Agency
WRA	Water Resources Act 1991

Chapter 1

AN OVERVIEW

INTRODUCTION

1.1 The law relating to contaminated land is a vast subject and this short handbook can only hope to provide a concise overview of it. It outlines the law, but also has regard to the practicalities behind the legal mechanisms: it is one thing to say that the law creates certain powers or duties, but quite another to assess how a particular legal regime will operate in practice. The political aspect is particularly important when one bears in mind the constant flow of environmental legislation coming from the EC. This is having a dramatic effect on all aspects of life and business in the UK, not least contaminated land.

To begin with, it is necessary to look at the ad hoc and pragmatic nature of the environmental law regimes within the UK.

FRAGMENTED REGIMES

1.2 Pollution can be transmitted through a variety of interconnecting environmental media and it is unlikely that any major incident will affect one medium only. Polluted air, for example, can contaminate the ground, which in turn may contaminate water. It would therefore seem logical for there to be a single regulatory body covering all environmental media, but this is not currently the case in the UK. For example, water pollution is primarily the responsibility of the National Rivers Authority (NRA), but air pollution is primarily the responsibility of Her Majesty's Inspectorate of Pollution and the local authorities. The position is further complicated by the common law: that is to say, decisions made by the High Court on private civil-law actions. The common law covers damage done to persons and property regardless of the particular environmental medium involved, but it does so in an uncertain and ad hoc fashion.

1.3 *Cambridge Water Co Ltd v Eastern Counties Leather plc* [1994] 1 All ER 53 involved the pollution of drinking water and addressed some fundamental principles of civil liability for damage done by pollution, causing great concern for businessmen as it did so. The facts and legal effect of this House of Lords judgment are discussed in Chapter 5, but, for the moment, I would note the chronology of the case. Cambridge Water sued Eastern Counties Leather for polluting a major drinking water supply. The case had major implications for businessmen, insurers and lenders, because it involved retrospective liability for historic pollution. In July 1991, Kennedy J found in favour of Eastern Counties Leather. Cambridge Water then appealed and, in

November 1992, the Court of Appeal overruled Kennedy J's judgment, holding that Eastern Counties Leather was liable to Cambridge Water for damages in the order of £1 million. Eastern Counties Leather, in turn, appealed to the House of Lords and, in December 1993, the House of Lords finally found in favour of Eastern Counties Leather. Thus, the law on this particular topic was in a state of flux for over 2 years, during the worst recession since the Second World War and at a time when legal uncertainty was the last thing the business community needed.

UNCERTAINTY OF STANDARDS AND GUIDELINES

1.4 Law and practice relating to contaminated land are still in their infancy, which is easily demonstrated by the fact that there are no universally understood meanings for concepts such as 'contaminated land', 'derelict land' and 'pollution' (see **2.2–2.13**). There are some fixed standards for particular regimes (eg drinking water), but they merely serve to highlight the paucity of standards in other areas.

1.5 While it is possible to look elsewhere in the world for guidance, foreign standards do not translate easily to the UK. Reference is often made to the standards used in Holland, but the geology of the Netherlands is markedly different from the geology of the UK, with the result that pollutants behave differently. Just as variations in geology can make it difficult to transport standards, so can differences in population density. A standard which produces an acceptable level of risk in a sparsely populated area may create an unacceptably high risk in a densely populated area.

1.6 The provision of standards is not a straightforward exercise but until some progress is made on that front, the practical application of the law is going to be a hit-or-miss affair depending on the ways in which individual regulatory bodies and the courts react to the diverse opinions of experts in each case. Given that background of uncertainty, in the UK one needs to try to understand what is demanded from regime to regime. This is a very difficult task. Take, for example, s 29 of the Environmental Protection Act 1990 (EPA), which provides that the Act covers any pollution capable of causing harm to man (including offence to his senses or harm to his property) or harm to any other living organisms supported by the environment (including harm to their health or other ecological systems of which they form part). This definition is not very helpful because its very breadth makes it difficult to predict how it will operate in practice. Thus, literally applied, killing a fly (or producing pollution which is capable of doing so) may fall within particular provisions of that Act.

1.7 In the absence of clear standards, s 29 will have to be applied, having regard to expert evidence, on a case-by-case basis and its meaning will be unravelled over time. Doubtless, lawyers and expert witnesses will enjoy the challenge of bringing definition to s 29, but its uncertainty does nothing to help those who have to apply it in the meantime, namely industrialists and those involved in property. One approach is to ask the relevant regulatory body what it expects from a particular industrial operation; however, even if the operator gets a clear answer from that body, he might still find himself prosecuted by an aggrieved neighbour or a pressure group who take a different view.

'Clean-up'

1.8 The phrase 'clean-up' is often used when describing legal and regulatory approaches to contaminated sites – however, it is misleading. Statutory regimes which require amelioration of pollution, such as s 161 of the Water Resources Act 1991 (WRA) (see Chapter 7), s 61 of the EPA (see Chapter 8) and the statutory nuisance provisions of the 1990 Act (see Chapter 9), are primarily designed to reduce the effects of pollution upon defined environmental media and affected land to acceptable levels, and that will often mean isolation of the contaminated material or subjecting it to treatment which reduces its pollutive effect – not removal of the material itself or neutralising its contamination. Private civil-law remedies (see Chapter 5) are confined to the parties to the litigation only and, in the case of contaminated land, the defendant will, at most, be obliged only to carry out the bare works necessary to isolate the plaintiff from the effects of the contaminant or to reduce those effects to levels acceptable to the courts.

1.9 Nor it is appropriate to speak of 'clean-up' requirements when a developer proposes to redevelop a contaminated site. The UK differs from some countries in its approach to remedial work on contaminated sites. In a number of countries (eg the USA, Canada and Holland) the developer must clean the site to its natural state or to a point where it can be used for any foreseeable purpose, whereas, in the UK, 'clean-up' is restricted to making the site suitable for a specified end-use. However, the fact that a site is fit for a particular end-use does not, by any means, make it 'clean': that end-use could be for a smelting or petroleum works. Thus a site may need to be cleaned on more than one occasion as it goes from use to use.

1.10 The phrase 'clean-up' is also misleading because it implies the possibility of a once-and-for-all solution in all cases, whereas the reality is that remedial action may have to be maintained over decades or even centuries. People use the phrase because it is comfortable and familiar. With reluctance, I have, for the reasons outlined above, tried to avoid using it in this book in favour of less misleading terms such as 'remedial measures', 'restoration measures' or 'preventative work'.

METHODS OF ENFORCEMENT

1.11 Each regulatory regime has its own particular methods of securing compliance, but they generally comprise a mixture of criminal and civil powers. For example, s 85 of the WRA makes it a criminal offence for any person to cause or knowingly permit the pollution of controlled waters.

Section 161 of the Act then gives the NRA the power to remedy the pollution and reclaim the costs of the remedial operation from the polluter. In general, regulatory bodies will be able to make directors, individual partners and other senior managers liable for the acts of their businesses. The director of a polluting company may find himself liable at criminal or civil law; in extreme cases, he may find he is disqualified from acting as a director for a number of years.

1.12 Criminal and civil liabilities may attach not only to the original polluter but also to those who are perceived as continuing the pollution. Thus a mortgagee may take possession of a contaminated site only to find himself liable for the pollution it generates. Indeed, in some cases, the lender may find that he has parallel liability as co-polluter where he has an active role in the conduct of the polluter's business, for example under a power of receivership. Also, lessees or tenants may find themselves liable for pollution started by the freeholder or a previous tenant.

1.13 In addition to purely legal powers, the regulatory bodies are well aware of the power of publicity. Indeed, one has to ask whether, given industry's desire to project a 'green' image, publicity is the most powerful weapon in any regulatory body's armoury. Many cases are settled by negotiation between polluter and enforcing body. The prospects of civil or criminal liabilities will focus the minds of both company and director, but financial penalties are often secondary to the image an organisation tries to project.

THE INTERNATIONAL DIMENSION

1.14 Many types of pollution do not recognise national boundaries, therefore no single country can properly formulate all of its policies for the environment in isolation from its neighbours. Furthermore, national environmental codes have a bearing on international trade, because national standards will differ in their effects on the costs of production or as inhibitions on import and export. As such, one has to have regard to the international dimension.

1.15 In particular, the UK must take account of its role in the EC. (The basis for EC involvement in UK environmental law and practice is complex and will be discussed at length in Chapter 4.) However, the international dimension does not end with the EC. Foreign businesses invest in the UK and UK businesses invest abroad, thereby exposing the country to foreign expectations about the environmental aspects of business transactions. For example, a businessman from the USA will be used to seeing environmental assessments when dealing in land in his country and will tend to be surprised to find that the same does not generally apply in the UK. If he is buying UK property, then he is likely to ask for an environmental assessment and thereby add to the knowledge and expectations of UK businessmen. The UK can also learn from the experiences of other countries. One example is the 'Superfund' regime in the USA, which was created with the objective of providing a comprehensive overriding fund for remedying pollution incidents in that country (see further **14.4**). Regardless of whether or not the 'Superfund' model could be applied to the UK, valuable lessons have been learned by studying how it operates in practice.

THE ROLE OF PRESSURE GROUPS

1.16 Pressure groups are also increasingly important. Organisations such as Greenpeace and Friends of the Earth, who are sophisticated and well advised, feature regularly in the courts: in May 1993 the *Lawyer* noted that Greenpeace was issuing three private prosecutions against ICI alleging pollution of coastal waters; on 15 March 1993 the *Independent* reported that the Anglers Co-operation Association was issuing private prosecutions against British Coal alleging the pollution of watercourses by closed mines; in September 1993 Greenpeace challenged British Nuclear Fuels in the High Court in a bid to prevent testing of the new nuclear reprocessing plant at Sellafield (THORP); then there was a challenge by Friends of the Earth over the Secretary of State's practice in enforcing water quality standards (see *R v Secretary of State for the Environment ex parte Friends of the Earth* (1994) *The Times*, 4 April). This pressure-group involvement should be a source of concern for directors and other senior executives, because both Greenpeace and Friends of the Earth have policies of issuing prosecutions against both company and director. A £20,000 fine would make little difference to a major multi-national, but it would probably make a dent in its chairman's bank balance.

MANAGING UNCERTAINTY

1.17 The law relating to contaminated land is currently characterised by its uncertainty, and this creates a considerable challenge for all those involved in land. This book demonstrates that vendor, purchaser, lender, end-user, corporate acquirer, insurer and receiver all need to be aware of potential liabilities because, when one deals in land, one also deals in the hidden risks which come with it. It is one thing to invest in land and make a proper return, but it is another to become involved in a 'bargain' site only to find it brings with it massive costs for remedying pollution from hidden contamination.

1.18 Persons involving themselves in property must assess risks before any hard commitments are made. Often, the costs of an environmental assessment will be tiny by comparison to overall project costs. A simple desk-top study would normally cost no more than £1,000 for a medium-sized site, yet this quick and simple exercise will give purchaser, vendor and lender valuable information about their proposed transactions. There must be an inter-disciplinary approach and cross-disciplinary awareness from the start of such projects. It is completely wrong for a developer to become involved in a project without ensuring that all areas of technical expertise are covered. For example, a hydro-geological survey will give information on the way in which the physical structure of the property will accommodate the proposed development; however, it will not normally tell the developer whether or not the land is polluted in any way or is capable of creating potential liabilities by cross-boundary pollution. These are matters which have to be taken into account at the very outset and before any form of contractual liability is incurred.

1.19 That does not mean that one should avoid all risk, because taking risks is what property investment is all about – but such risks should be calculated ones based on knowledge of the site and its potential liabilities. Nor does it mean that one has to carry out all relevant site surveys and field tests beforehand. Every transaction has its own dynamics and it will often not be possible to support a heavy cash outlay at the very start of a project. A development may depend on obtaining planning permission and, as such, it might be unwise to invest large sums of money until the principle of development is accepted by the local planning authority. With a large project, precise details of layout and siting might not be settled for months, years or even decades: it would, in such a case, be impractical and extremely unwise to engage in full detailed site surveys for contamination at the start of the development. Those involved in the transaction should, however, take such steps as are prudent and financially sound at the outset. They should also formulate a strategy which ensures that detailed studies are carried out as the project progresses.

ECO-MANAGEMENT AND AUDIT

1.20 As discussed above, one defence strategy is to ensure, as far as possible, that the site carries no unacceptable pollution risk before completing the transaction – but that strategy does not guard against post-transactional pollution. It would be unfortunate if a lender advanced money on the security of a 'clean' site only to find, when he enforced that security, that the borrower had contaminated the site. The lender may have been wise enough to get indemnities, warranties or guarantees in the hope of covering such an eventuality, but these undertakings may turn out to be either worthless or exceedingly troublesome to enforce. The prudent lender will, therefore, keep the borrower 'green' by specifying in the loan agreement that the borrower's operation must comply with specified environmental criteria. In particular, the lender may require the borrower to undergo regular environmental audits and, in some cases, follow environmental management regimes. It may be necessary to require certain borrowers to enter the BS 7750 or Eco-Management and Audit regimes. Likewise, lessors or head-lessees may find it appropriate to require their tenants or subtenants to comply with similar environmental audit/management regimes.

Chapter 2

WHAT IS CONTAMINATED LAND?

INTRODUCTION

2.1 This is a very short chapter examining some of the basic terminology which will be used throughout this book. One finds that, even at this level, there are problems in obtaining clear definitions of words and phrases.

DEFINING THE CONCEPTS

'Derelict land'

2.2 It is generally agreed that 'derelict land' has a wider meaning than 'contaminated land'. Land may be derelict without being contaminated: for example, the cause of dereliction may be that the site is covered by old buildings or heaps of inert spoil. The Derelict Land Act 1982 gives the Secretary of State for the Environment powers to advance derelict land grants for land which is 'derelict, neglected or unsightly', but it does not define the word 'derelict'. However, the Department of the Environment (DoE) has produced its own definition for the purposes of derelict land grants:

> 'Land so damaged by industrial or other development that it is incapable of beneficial use without treatment'. (*Derelict Land Grant Advice Note 1* (May 1991), para 18.)

That definition is repeated in para 5.68 of *The UK Environment* (DoE, 1992).

'Contaminated' and 'polluted' land

Distinctions

2.3 Many commentators have sought to distinguish between 'contamination' and 'pollution' on the basis that calling a site 'contaminated' is a statement about its general condition without saying whether it is harmful or not, whereas 'pollution' implies the potential for harm to man or the environment. In the consultation paper 'Public Registers of Land which may be Contaminated' (May 1991), the DoE stated:

> '3.1 The government share the view of the Royal Commission on Environmental Pollution that contamination is not synonymous with pollution. The Commission regarded contamination as "a necessary but not sufficient condition for pollution", and also noted that "what today is regarded as no more than contamination may tomorrow be pollution, as a substance accumulates to … harmful concentrations, or as knowledge is gained of hitherto unsuspected effects".

> 3.2 The Commission therefore used "contamination" to denote the presence in the environment of certain substances, without making a judgment on whether they are actually harmful. The use of the term in this paper follows this principle.'

2.4 In its draft Code of Practice DD175: 1988 (the 'Code of Practice for the Identification of Potentially Contaminated Land and its Investigation'), the British Standards Institute (BSI) states, at p 4:

> 'The presence of contaminants on a site does not necessarily mean that a hazard exists, or that it is essential for action to be taken. Many contaminated sites would present no hazards to their users or occupiers because the uses are "tolerant" of the contamination.'

On this approach, merely asking whether or not land is 'contaminated' is not, by itself, a worthwhile exercise because all land in the UK is, in all probability, contaminated in the sense of being subject to substances not naturally arising there, and even the highest mountain-top will have been subject to airborne pollution.

The notion of harm

2.5 Generally, 'pollution' tends to involve the notion of harm. In relation to controlled waste, s 29(3) of the Environmental Protection Act 1990 defines 'pollution of the environment', in effect, as the release or escape to land, water or air of substances or articles constituting or resulting from such waste and capable (by reason of the quantity or concentration involved) of causing harm to man or any other living organisms supported by the environment. It is important to stress that, on this interpretation, 'pollution' does not necessarily equate with immediate identifiable environmental harm. Section 29(3) of the 1990 Act is cast in terms of *potential* harm, that is the release or escape of a thing *capable* of harming man or the environment. Proving potential harm in a criminal or civil case is different from (but not necessarily easier than) proving actual harm. That focus on capability or potential is supported by *National Rivers Authority v Eggar (UK) Ltd* (1992) 4 LMELR 130, where the defendant company was found guilty of causing 'polluting matter' to enter the River Tyne in breach of s 85 of the Water Resources Act 1991. The Crown Court held that the offence was committed if the run-off was capable of causing harm to the extent that it may damage the potential usefulness of the river.

Section 29(5) then goes on to define 'harm' as harm to the health of living organisms or other interference with the ecological systems of which they form part and, in the case of man, includes offence to any of his senses or harm to his property.

2.6 The position is unfortunately complicated by the fact that 'contamination' is often used to denote harm. The Environment Act 1995 gives the following definition (see **16.6**):

> '"Contaminated land" is any land which appears to the local authority in whose area it is situated to be in such a condition, by reason of substances in, on and under the land, that:
>
> (a) harm, or
> (b) pollution of controlled waters
>
> is being, or is likely to be caused.'

The 'suitable for use' approach

2.7 The propensity of a site to cause harm is bound to be related to its use. For example, if the site is hard-surfaced and used for a car-park, its propensity for causing harm may be different to where it is used for allotments. A landfill site may be venting explosive landfill gases in safety, but danger of explosion may occur if buildings are constructed on it or nearby, because the buildings may trap the gases.

2.8 Draft BSI Code of Practice DD175: 1988 states (at para 10.1, p 21):

> 'When assessing the suitability of a contaminated site for development, the central principle to be adopted is that the eventual users or occupiers of the site, buildings or building services, and plants and animals, ought not to be subjected to risks which could be avoided.'

This 'suitable for use' approach is the one adopted by the DoE in relation to the redevelopment of contaminated land: see **11.8**.

Clarifying the terminology

2.9 Some definitions sit uneasily on the fence. The definition coined by the NATO Committee on Challenges to Modern Society is:

> 'Land which contains substances which, when present in sufficient quantities or concentrations, are likely to cause harm, directly or indirectly, to man, to the environment, or on occasion to other targets.'

Whilst this NATO definition has been cited by some commentators, it is not clear as to whether or not it is referring to substances which currently pose a threat of harm or to substances which do not currently pose a threat but could do so in the future if present in higher quantities or concentrations.

2.10 It is, therefore, incumbent on those tackling the subject of contaminated land to set out their terminology at the outset. Throughout this book, the word 'pollution' is used to mean:

> 'the release or escape to land, water or air of any substance, liquid or gas which is capable of causing harm to the health of any living organism or other interference with the ecological systems of which they form part and, in the case of man, includes offence to any of his senses or harm to his property.'

This is generally consistent with modern usage.

2.11 However, for the purposes of Chapter 14 and the model documents in Appendix 1 it has been necessary to use a wider definition of 'contamination' which does not involve assessing harm or potential harm, namely:

> 'the presence on, in or under any land, or on or in water, or in the air of any substance, liquid or gas not naturally occurring there.'

It has been necessary to use this wide definition to ensure certainty in the drafting of preliminary inquiries, indemnities, warranties and so on, because, if a requirement to show harm or potential harm were made, then there would be too much room for debate as to whether or not a particular inquiry, indemnity, warranty and so on applies to a given site or situation. If the recipient of a preliminary inquiry is requested to give details of all contamination (ie all substances not naturally occurring there) then he may have an onerous task, but he will have very little room for obfuscation. Furthermore, this definition encompasses situations where liability may arise when the quality of an environmental medium is diminished but not necessarily harmed: see the discussion of *National Rivers Authority v Dovermoss Ltd* and *National Rivers Authority v Egger (UK) Ltd* at **7.7–7.9**.

2.12 Separating 'contamination' and 'pollution' in this way is not an abstract academic exercise: the practical difference between the terms can be seen in the drafting techniques discussed in Chapter 14. Next, one has to translate these descriptions and definitions into standards or guidelines which can be measured and applied on a day-to-day basis in real life – that topic is tackled in Chapter 13.

2.13 It is important to stress, at this early point, that the concepts of 'contamination' and 'pollution' are variable and relative. They are variable in that their content will change as opinions or value judgements on environmental impact change. They are also relative to the ability of experts to detect and measure environmental impacts – and, as Chapter 13 will show, that is a far from exact science.

SOME TYPICAL RISKS

2.14 Pollution from contaminated land can pose risks to the health and safety of people living near it. For example, in 1985 a methane build-up from a landfill caused an explosion at Loscoe in Derbyshire which wrecked a bungalow. On 19 May 1993 evidence of substantial petrol pollution from leaking service-station tanks was mentioned in a House of Commons debate. During the debate, Mike Hall, MP, observed that, in a recent incident, 17,000 gallons of petrol were recovered from around a Shell site at Longford Bridge in Warrington. He also noted that 20–25,000 gallons of petrol had found their way into the ground around a service station in Braintree. In Scarisbrick, Lancashire, a petrol leakage affected 20,000 square metres of land and remedies included pumping gallons of petrol from domestic cellars – with 150 gallons being pumped from one home. Alan Watson reports (*Ecologist*, vol 23, no 5, pp 174–8) that a maternity hospital and at least four food-related factories were built on the site of a former chemical works – the land being affected by the carcinogenic hexavalent chromium. Many of the affected areas were children's play areas and football pitches where some of the toxic waste glowed at night.

2.15 Pollution from contaminated land can also affect watercourses and groundwater, and a high proportion of our drinking water comes from boreholes which draw upon groundwater resources. Rightly, there has been widespread concern about groundwater pollution in this context. In the 1970s a tannery in Sawston, Cambridgeshire, caused solvent contamination to a source of a drinking-water supply to 275,000 people. The Cambridge Water Company lost almost £1 million in remedial action and sued the tannery (see **5.22–5.29**: *Cambridge Water Co Ltd v Eastern Counties Leather plc* [1994] 1 All ER 53). Watson (above) reports that a borehole in Blewbury, Oxfordshire, was closed in 1992 because of pollution by carbon tetrachloride, the only possible source being the Atomic Energy Authority (AEA) laboratories at nearby Harwell. The AEA is reported to be committed to expenditure in remedial actions running into millions of pounds over a decade.

2.16 Abandoned mines are another major problem. The National Rivers Authority (NRA) report entitled *Abandoned Mines and the Water Environment* (Water Quality Series No 14) states, in relation to abandoned coal mines, at para 4.3 that 'Close to 100 discharges, mostly originating from underground workings, are currently causing considerable concern. Some 200 km of rivers, streams or brooks are affected'. The report contains a list of affected watercourses, which is reproduced in Appendix 2(1). In respect of abandoned metal mines, the report notes that some 400 km of waters are affected. The principal problem area is the South-West, with about 1,700 abandoned mines in the region – including the infamous Wheal Jane mine, from which millions of gallons of contaminated water escaped into the Carnon River and Restronguet Creek in January 1992 causing serious contamination problems and involving the NRA in long-term remedial measures costing millions of pounds. The Environment Act 1995 provides new powers designed to overcome this problem with mines abandoned after 31 December 1999: see **16.56**.

2.17 Nor are contamination problems confined to urban or 'brown-field' sites. In strict terms, one is unlikely to find a truly 'green-field' site anywhere in the UK. In 1994 the ENDS Report 228, p 10, noted that concentrations of agricultural pesticides in streams draining arable areas have been tested as regularly above EC drinking-water limits. Even the apparently innocuous practice of dipping sheep can cause contamination: see ENDS Report 230, p 11. Other studies have raised concern about the environmental effects of spreading sewage sludge on agricultural land, particularly in relation to the effects of metal-rich sludges on soil ecosystems: see ENDS Report 229, p 14, and ENDS Report 232, pp 12–13. Watson (above) reports that the use of dichloropropene-dichloropropane to kill worms on agricultural land has, in combination with other pesticides, so polluted 30,000 hectares of agricultural land in the Netherlands that local water supplies may need to be treated until the year 2050: see also the NRA report *The Influence of Agriculture on the Quality of Natural Waters in England and Wales* (1992).

Chapter 3

SOURCES OF LAW AND PRACTICE

INTRODUCTION

3.1 What is sometimes simplistically described as 'Environmental Law' is, in fact, a complex inter-relationship between law and administrative practice. It is essential to understand not only the elements of the structure but also how it all fits together – or sometimes does not! This chapter will discuss the underlying legal mechanisms and then go on to discuss the way in which the regulatory bodies and ministries of state overlay this framework with a considerable and very important edifice of administrative and bureaucratic practice. The common law, which is a very important source of law, is discussed in Chapter 5 (save for judicial review which is discussed below).

LEGISLATION

3.2 The fundamental source of most UK law is in Acts of Parliament, which override all other sources of law. This concept is known as the 'sovereignty of Parliament' although it often appears to be somewhat strained by the impact of law from the EC: see Chapter 4.

3.3 Acts of Parliament are often supplemented by delegated or 'secondary' legislation, including statutory instruments, regulations and bye-laws. Delegated legislation is not made by Parliament, but by ministers, government departments and other bodies. Much environmental law depends on a combination of Acts of Parliament and delegated legislation. Many statutes are 'framework' legislation and only become fully effective when matters of detail are provided by delegated legislation. The relevant Act of Parliament may state that its provisions can be supplemented or limited by delegated legislation. Thus, whilst s 33(1) of the Environmental Protection Act 1990 (EPA) sets out general requirements in relation to the management of controlled waste, s 34(3) provides that the Secretary of State for the Environment may disapply those provisions by regulations. The position has been very heavily modified by the Waste Management Licensing Regulations 1994 (SI 1994/1056).

The use of delegated legislation saves valuable Parliamentary time on matters of detail, but it arrogates considerable power to the body or person enacting it, and is seen by many as being no better than a necessary evil.

3.4 There are a number of restrictions on the way in which these bodies may use their delegated legislative powers. First, they are normally entrusted with the power to create delegated legislation by Acts of Parliament, and, as such, that delegated legislation must be produced strictly within the terms laid down by those enabling Acts. A statutory instrument which falls outside the powers conferred by the relevant Act may be struck down by the High Court as being 'ultra vires' (ie beyond the powers) of the Act. Secondly, many statutory instruments are laid before Parliament before becoming effective and may be challenged by any Member of Parliament during that period. Parliament has a select committee which may scrutinise delegated legislation which is considered oppressive. Also, some types of delegated legislation cannot be brought into effect until a public inquiry procedure has been followed. Ultimately, Parliament may, of course, revoke a misused delegated power.

The problem of deferred commencement

3.5 An Act may provide that certain of its provisions are not effective immediately but will be brought into force by statutory instrument. Thus, although the EPA received the Royal Assent on 1 November 1990, most of Part II of the Act did not come into force until 1 May 1994, when the appropriate statutory instrument produced by the Department of the Environment (DoE) took effect.

3.6 Some provisions are never brought into force. Section 143 of the EPA empowered the Secretary of State for the Environment to pass regulations to create registers of contaminative uses of land; however, the Secretary of State never used the power and s 143 has now been repealed by the Environment Act 1995. Section 1 of the Control of Pollution Act 1974 (COPA) imposed a duty on waste-disposal authorities to ensure that arrangements for the disposal of waste within their areas were adequate, yet that section has never been brought into force – despite the fact that the COPA received the Royal Assent on 31 July 1974.

3.7 Whilst deferred commencement dates serve a useful purpose in allowing government departments time to prepare matters of detail after the main legislation has been passed, it has the disadvantage of creating uncertainty. The deferred commencement provision also gives the government breathing space to change its mind. Indeed, it could be argued that a cynical government might pass legislation in order to make a political gain yet with no real intention of ever bringing it into force. Alternatively, of course, there could be a general election and change of government before the legislation can be brought into force.

THE COURTS

3.8 This section will be confined to outlining the civil and administrative courts system; the criminal courts are discussed in Chapter 6. For present purposes, the civil and administrative courts can be treated as being:

- **magistrates' courts**, dealing with statutory appeals (eg under s 80(2) of the EPA) or administrative matters (eg nuisance orders under s 82(2) of the EPA);
- **county courts**, dealing with less complex cases including the recovery of statutory expenses (eg s 81(4) of the EPA); and
- **the High Court**, dealing with more complex cases and exercising specific control over regulatory bodies.

3.9 The High Court of Justice (or 'High Court' for short) is the lower branch of the Supreme Court of Judicature (the higher branch being the Court of Appeal). The functions of the court which are important in the present context are:

- the issue of injunctions to enforce regulatory requirements – for example the issue of injunctions under s 24 of the EPA in respect of integrated pollution control registers and under s 81(5) of the EPA in respect of statutory nuisance;
- hearing appeals arising from the decisions of county courts, magistrates' courts and certain tribunals; and
- deciding applications for the 'judicial review' of the actions of certain tribunals, courts and other bodies.

Judicial review

3.10 'Judicial review' is of considerable importance in the present context. The bodies which provide environmental regulation derive all of their powers from, and are often created by, statute. For example, the National Rivers Authority (NRA) was created by the Water Act 1989 and its regulatory powers are currently prescribed by the Water Resources Act 1991. 'Statutory nuisance' (see Chapter 9) is controlled by the local authorities and derives entirely from Part III of the EPA. A regulatory body, such as the NRA or a local authority, is not allowed to go beyond the powers and duties laid down for it by statute. If it does so, then it is said to be acting 'ultra vires' (ie beyond its powers) and its action can be struck down (or 'quashed') by the High Court. If the relevant statute provides a mechanism for a High Court challenge on the ground of ultra vires, then that mechanism must be used, but if the statute does not, then the High Court may, on the application of a person aggrieved, judicially review the action or inaction of the relevant body.

Statutory discretion

3.11 Unfortunately, the position is complicated by the fact that many statutes give wide discretions to regulatory bodies. For example, s 35 of the EPA allows the waste-regulation authorities to attach such conditions as appear to them to be appropriate to waste-management licences. In another context, s 143 of the EPA stated that the Secretary of State 'may' set up registers of contaminative uses of land, but did not require that he 'must'. The fundamental reason for giving a person or body a wide statutory discretion is to allow him or it to get on with the appointed job as to the matters of detail, and to allow him or it to exercise political or professional judgements in the discharge of that task. If the statute had intended that judges be allowed to intervene, then it could easily have said so. Notwithstanding these arguments, judges have often intervened in the exercise of statutory discretions – although in doing so they run the risk of being accused of interfering in political matters, despite being unelected officials drawn from a very narrow social and professional background.

'Unreasonableness' and procedural error

3.12 The High Court has, in order to challenge statutory discretions, extended the notion of 'ultra vires' to include 'unreasonable' behaviour. The notion of 'unreasonableness' is a very narrow legal one and has nothing in common with the ordinary use of the word. The classic formulation, known as the '*Wednesbury* principle', was stated by Lord Greene MR in *Associated Provincial Picture Houses Ltd v Wednesbury Corporation* [1948] 1 KB 223:

> 'Lawyers familiar with the phraseology commonly used in relation to exercise of statutory discretions often use the word "unreasonable" in a rather comprehensive sense. It is frequently used as a general description of the things that must not be done. For instance, a person entrusted with a discretion must direct himself properly in law. He must call his own attention to the matters which he is bound to consider. He must exclude from his consideration matters which are irrelevant to the matter he has to consider. If he does not obey those rules, he may truly be said, and often is said, to be acting "unreasonably".'

3.13 The High Court will also intervene where a decision-maker has reached his decision by means of a tribunal, hearing or other administrative procedure involving the participation of those affected by the decision and has, in doing so, committed a procedural error. This type of intervention is often described as the 'right to a fair hearing' (or, somewhat philosophically, 'the principle of natural justice') and requires that persons involved have the right to hear, and comment on, the case against them and that the decision-making process is not biased.

3.14 It is necessary to stress that judicial review does not purport to examine the 'merits' of the case (ie matters of fact, opinion and judgement), merely the legality of the disputed decision or action. The court does not try to stand in the shoes of the decision-maker and remake his decision for him. Indeed, many decisions are simply referred back to the original decision-makers with the requirement that they carry out the exercise again in accordance with the law. Nevertheless, the reality is that any judgment based on a perception of what is 'reasonable' (however narrowly the term is defined) will draw the judge into the merits and politics of the case. This point is of particular importance in the context of environmental law, where nothing is self-evident and, it seems, everything is a political football.

Remedies

3.15 If an application for judicial review is successful, then the High Court may use one or more of the following remedies:

- an injunction (see **5.44** for an explanation of this remedy);
- a declaration – a statement as to the legal position between the parties which does not actually change their respective rights and obligations (although, in practice, most public bodies will act in accordance with a declaration);
- certiorari – an order quashing the disputed decision or action completely, including any relevant orders, notices or other documents;
- prohibition – an order stopping the execution or continuance of the disputed decision or action; and
- mandamus – an order requiring an administrative or regulatory body or person to carry out a stipulated action.

Who is judicial review available to?

3.16 Judicial review is available only to persons who can show a 'sufficient interest' in the disputed decision or action. This is relatively straightforward where the applicant is directly involved in the subject of the application (ie having been an objector at a public inquiry), but it becomes complicated where the applicant's connection is indirect or the applicants are a pressure group objecting on moral, social or political grounds. In *R v Secretary of State for the Environment ex parte Rose Theatre Trust Company* [1990] 1 All ER 754, a group concerned as to proposals to develop the site of the Rose Theatre in Southwark were refused leave to pursue a judicial review of the Secretary of State's refusal to list the site as an ancient monument. In effect, Schiemann J took the view that the wording of the relevant legislation precluded such a challenge, because the Secretary of State's power had to be exercised on behalf of the public as a whole as opposed to any particular group or person. In *R v Poole Borough Council ex parte Beebee* [1991] JPL 643, the British Herpetological Society was allowed to pursue the judicial review of proposals to build houses on a site of special scientific interest, because it had actually been involved in the site in the past. However, the World Wide Fund for Nature (who financed the action) were said *not* to have a sufficient interest, whereas the Nature Conservancy Council did because it had specific duties in relation to nature conservation. A pressure group without a sufficient interest can, in practice, get around these problems by funding an action in the name of a person who does have sufficient interest.

3.17 In 1993, the High Court held that Greenpeace had sufficient interest to challenge the grant of a licence 'variation' by Her Majesty's Inspectorate of Pollution (HMIP) allowing British Nuclear Fuels (BNF) to conduct tests for a nuclear reprocessing plant (THORP) in Cumbria: *R v Her Majesty's Inspectorate of Pollution ex parte Greenpeace* [1994] 2 CMLR 548 (see ENDS Report 225, pp 42–3). The construction of the plant had been allowed following the Windscale public inquiry in 1977. Otton J stated that Greenpeace was 'an entirely responsible and respected body with a genuine concern for the environment', with 2,500 members in Cumbria and a bona fide interest in the activities of BNF. He noted that, without the involvement of Greenpeace, concerned individuals would not have the resources or expertise to present their cases in an effective manner. In the event, Greenpeace was not successful in its challenge: see *R v Secretary of State for the Environment ex parte Greenpeace and Lancashire County Council* (1994) *The Independent*, 8 March; ENDS Report 230, p 43.

3.18 However, it is notable that Greenpeace were not required to pay the legal costs of the Government and BNF. This marks a departure from the usual principle that the losing party meets the winner's legal costs and was justified on the basis that the case was of substantial public interest, that leave had been granted and that Greenpeace had succeeded on one important point relating to EC law. This judicial approach to costs is reflective of a provisional recommendation of the Law Commission in 1993, that the usual civil litigation approach to costs should not apply to cases brought in good faith in the public interest. Thus, whilst Greenpeace were unsuccessful in stopping the THORP proposals, they were successful in consolidating the rules relating to sufficiency of interest and costs. This success will encourage greater pressure-group involvement in public-interest litigation.

THE ADMINISTRATIVE FRAMEWORK

Regulations and guidance

3.19 Whilst Acts of Parliament vest the various regulatory bodies with their powers and duties, those statutes are often skeletal and will often be largely meaningless when read in isolation. If one is considering, say, the regulatory matters which need to be taken into account before starting a development project, then it is necessary to look not only at the statute but also at the statutory instruments produced by the relevant ministry and relevant 'informal' guidance and advice. For example, if the project involves the transportation and deposit of waste material, it is necessary to have regard not only to the EPA but also to the Waste Management Licensing Regulations 1994 (SI 1994/1056), the Environmental Protection (Duty of Care) Regulations 1991 (SI 1991/2839), and DoE Circulars 11/94 and 19/91 in determining whether the waste is subject to the 'duty of care' under the EPA and whether the activities need a waste-management licence: see Chapter 8.

3.20 The need to look at relevant regulations is self-evident; however, the important point here is that the regulations will be made by an administrative or regulatory body. Thus, it is also necessary to consider that body's understanding of its own regulations and any proposals it may have for amending or replacing those regulations. Many regulatory bodies act by way of discretionary powers, as opposed to mandatory duties, or have to contend with ambiguous legislation. They often provide written advice to clarify the ways in which they will act and it is important to take it into account.

3.21 Government departments often produce the most authoritative written guidance, because they often act as bodies of appeal for the decisions of regulatory bodies. For example, the Secretary of State for the Environment and the Welsh Office decide appeals from refusals of planning permission by local planning authorities and issue copious guidance on all aspects of the planning regime in 'planning policy guidance notes', circulars, consultation papers and even ministerial speeches: see Chapter 10. The decisions of the Secretary of State or his inspectors on past planning appeals are also used to provide illumination and authority. Likewise, the Secretary of State for the Environment discharges numerous functions under the EPA, including supervising and hearing appeals from HMIP, waste-regulation authorities and local authorities. Again, he has issued copious guidance in the form of waste-management papers, circulars and so on.

Some ministerial codes of practice are given statutory recognition, for example s 61E of the Land Drainage Act 1991, and s 34(7) and (10) of the EPA.

Administrative remedies

3.22 Those aggrieved by administrative decisions or actions often cannot afford to mount High Court challenges and, even if they have the resources, the courts have only limited jurisdictions and powers of review and might not be able to help. Maladministration by a public body is not, by itself, enough to justify action in the courts: see *R v Knowsley Borough Council ex parte Maguire* (1992) 90 LGR 653. In such cases, it might be possible to have recourse to the 'Ombudsman'.

3.23 If the offending body is a department of central government (eg the Department of Transport, the DoE etc) then a complaint may be made to the Parliamentary Commissioner for Administration (PCA) under the Parliamentary Commissioner Act 1967. If the complaint is in relation to a local authority, or the NRA, then the correct body is the Commission for Local Administration under the Local Government Act 1974. There are two separate commissions for England and Wales. The commission for England has divided England into areas and appointed one or more local commissioners for each area. Complaints go to the local commissioners.

It is not possible to complain directly to the PCA or a Local Commissioner for Administration (LCA) – such complaints must be via an MP or local councillor, although an LCA may investigate a complaint where a member of the authority has been requested to refer a complaint to the LCA but has failed to do so.

Matters excluded from investigation

3.24 Both the Parliamentary Commissioner Act 1967 and the Local Government Act 1974 set out lists of matters which are excluded from investigation and they need to be examined before any complaint is made. In general, neither the PCA nor an LCA may investigate a complaint made more than 12 months after the complainant had notice of the matters alleged in the complaint, but the period can be extended in special circumstances. Importantly, neither may investigate where the complainant has a remedy by appeal, reference or review or, in a court of law, unless satisfied that it would be unreasonable to expect him to resort to it.

3.25 In a 1990 complaint against Macclesfield Borough Council, an LCA found that there was maladministration and consequent injustice to the complainants, the local Chamber of Commerce, because the council had failed to carry out adequate public consultation before making a road-closure order. The Chamber could have challenged the order in the High Court but did not do so. The LCA decided to investigate, notwithstanding that point, because of the public interest component and because it would have been unreasonable to expect the Chamber to bear the costs of legal proceedings: [1992] JPL 577.

In a 1988 complaint against Gloucestershire County Council, an LCA found that there had been maladministration in relation to the enforcement of planning controls in respect of a quarry and that the complainants had suffered injustice by reason of disturbance and annoyance from noise, dust and lorry traffic. The point of interest, in the present context, is that the complainants appear to have had a prima facie case for an action in the courts founded on nuisance and, perhaps, trespass. It is not clear whether the council raised this point, but it is reasonably apparent that the possibility of a civil claim did not deter the LCA: see [1994] JPL 69; also see [1993] JPL 691 and [1993] JPL 695.

3.26 Neither the PCA nor an LCA is entitled to investigate the commencement or conduct of civil or criminal proceedings or action taken in the investigation of crime. Given that a large number of pollution matters involve potential civil or criminal liability, this restriction is important in the current context.

3.27 In practice, some administrative bodies raise these legal restrictions to prevent investigation of their conduct, but some bodies take the attitude that they have nothing to hide and do not try to shelter behind the restrictions. The prudent complainant will, in the case of doubt, try to engender the 'right' attitude in the relevant body by media coverage and the lobbying of its members and officers.

Function and powers of the PCA and LCA

3.28 Whilst the PCA and the LCA will consider the legal background to any complaint, their primary function is to ascertain whether or not there has been any 'maladministration' insofar as the body has failed to act in accordance with certain standards of administrative conduct and whether, in consequence, the complainant has suffered injustice.

3.29 Neither the PCA nor the LCA have any powers to force the bodies they investigate to remedy any complaints or pay compensation; however, their recommendations are often followed – especially if accompanied by media interest.

Chapter 4

THE EUROPEAN DIMENSION

INTRODUCTION

4.1 In 1946 Winston Churchill proposed a 'sovereign remedy' for the conflicts which had divided Europe during the first half of this century, namely a 'United States of Europe'. That sovereign remedy has evolved into a complex series of ties between European states, existing within a framework now called the European Community (formerly called the European Economic Community). Regardless of the way in which its institutions have expressed their policies and decisions, the EC has never been a purely economic union. It has always had social and political functions and, latterly, has become involved in the environmental controls operated by Member States.

4.2 Whether we like it or not, the European influence permeates all aspects of UK commercial and environmental law and practice. This chapter does not attempt to give an overview of all the manifestations of that influence, because that would be a massive task. Instead, it provides a very elemental thumb-nail sketch of some relevant European systems and institutions, with particular reference to the EC, and then outlines the broad concepts which lie behind EC and European environmental policies and legislation. This is not, even at this level, easy reading; however, it is essential in order to follow the detailed discussion in later chapters.

THE THREE 'COMMUNITIES'

4.3 Churchill's sovereign remedy has, in part, evolved into three distinct but interlinked 'communities', namely:

- the European Coal and Steel Community;
- the European Community of Atomic Energy; and
- the European Community.

The European Coal and Steel Community (ECSC)

4.4 The ECSC was established in 1951 by the Treaty of Paris, with the primary objective of tying France and Germany together in the production and distribution of coal and steel. The original ECSC was joined by France, Germany, Italy, Belgium, the Netherlands and Luxembourg. The Treaty of Paris is due to expire in the year 2001 when the matters covered by it will then fall within the control of the EC.

The European Atomic Energy Community (Euratom)

4.5 The Euratom was formed in 1957 and the Euratom Treaty was signed at the same time as the Treaty of Rome. Euratom is concerned with the promotion of the peaceful and constructive use of atomic energy in Europe.

The European Community

4.6 The Treaty of Rome (sometimes called the EEC Treaty) was signed on 25 March 1957 and formed the basis for what was then known as the European Economic Community. The UK joined the EEC and the other two communities on 1 January 1973. The Treaty of Rome has been supplemented by the Single European Act 1986 and the Maastricht Treaty (also called the Treaty of European Union) and, with the theoretical completion of the Single Market in 1992, the term 'European Economic Community' has been changed to the 'European Community'. It currently comprises 15 Member States: namely Belgium, Denmark, France, Germany, Greece, Ireland, Italy, Luxembourg, the Netherlands, Portugal, Spain, the UK and, since 1 January 1995, Austria, Finland and Sweden.

4.7 The Treaty of Rome set out four broad economic objectives:

- harmonious development of economic activities;
- continuous and balanced expansion;
- an increase in stability; and
- an accelerated raising of standards of living.

These broad objectives were pursued by particular provisions which sought:

- freedom of movement between the Member States for goods, persons, services and capital;
- freedom of competition between businesses within the Member States;
- a common trading policy with countries outside the EC;
- common policies for agriculture and transport; and
- the harmonisation of national laws and co-ordination of national economic policies.

The Treaty also set out political and social objectives including encouraging closer union among the peoples of Europe, and promoting social and regional policies.

The merger of European institutions

4.8 A Treaty of Merger was signed in 1965 and provided that, from 1967, the main institutions of the three communities would be merged to give one Council of Ministers, one Assembly (ie the European Parliament) and one Court of Justice to serve all three. This merger was at an administrative level and, for the moment, each of the communities still retains its separate legal identity and capacity, identified by its own discrete treaties, legislation and case-law.

THE SINGLE EUROPEAN ACT 1986

4.9 The Single European Act was ratified in 1986 and came into force on 1 July 1987. Its primary function was to remove trade barriers between Member States and bring about the Single European Market by the end of 1992. One of the main objectives of the EC's '1992' policy was to promote a level playing field, in terms of production costs, across the Community, which involved the harmonisation of standards between Member States, including those relating to the environment, health and safety and consumer protection. However, the Act also included a number of fundamental environmental policies, including the concept of sustainable growth, and provided changes in the balance of powers between the institutions of the EC.

THE MAASTRICHT TREATY

4.10 The Treaty of European Union ('the Maastricht Treaty') came into effect on 1 November 1993 and promotes five broad concepts:

- the European Union, to be comprised of the current EC, a common foreign and security policy, and an inter-governmental home affairs and justice policy;
- the concept of 'subsidiarity', whereby the EC institutions will limit their decisions only to those matters which are strictly necessary and, so far as is compatible with efficiency and democracy, allow decisions to be made by national, regional or local government where possible;
- a Committee of the Regions to offer advice in relation to the regions of Europe;
- the principles of economic and monetary union; and
- European citizenship.

The Treaty also adjusts the balance of power between the Community's institutions, including enhancing the powers of the European Parliament.

4.11 The terms 'European Union' and 'European Community' are now often used interchangeably – in many cases it is as though the EC has somehow been absorbed without trace into the new regime. In fact, the EC remains in existence as a discrete legal system and remains the main foundation of Community law. Accordingly, 'European Community' and 'EC' are the preferred terms used in this book.

THE INSTITUTIONS OF THE EUROPEAN COMMUNITY

The European Commission

4.12 The European Commission can be described as the bureaucratic and executive arm of the EC. Its overriding duty under the Treaty of Rome is to see that the provisions of Community treaties and the decisions of Community institutions are properly implemented. The Commission has been called the 'watchdog of the Community'. If negotiations fail, it can enforce

EC requirements through the European Court of Justice and, by virtue of the Maastricht Treaty, ask that the offending state be fined.

One of the most important functions of the Commission is to initiate EC secondary legislation and ensure that the legislation has been carried out. It is important to recognise that, whilst the EC Council can adopt EC secondary legislation, only the Commission can initiate it.

4.13 The Commission is divided into 23 Directorate-Generals of which DGXI – 'Environment, Nuclear Safety and Civil Protection' – is of particular importance to environmental matters. Each Directorate-General is headed by a Commissioner and is divided into separate Directorates. Directorate-General DGXI is divided into three Directorates.

The European Parliament

4.14 Since 1994, the Parliament consists of 626 members (MEPs), directly elected for 5 years by each state using its own national electoral system. The number of MEPs elected to the Parliament from each Member State is related to the population size of that State: for example, Germany has 99 MEPs, France, Italy and the UK each have 87 MEPs, whereas Luxembourg (the smallest state) has 6. The Parliament is presided over by a president, who is supported by 14 vice-presidents.

4.15 The European Parliament was established to exercise 'advisory and supervisory powers' conferred on it by the Treaty of Rome. The Parliament does not have powers to initiate EC legislation (that is restricted to the Commission only); however, it has power to withhold the Community budget and, in extremis, to force the resignation of the Commission. The Single European Act 1986 enhanced the Parliament's ability to amend legislation, and its power has been further increased by the 'co-decision' procedure introduced by the Maastricht Treaty.

The Council of Ministers

4.16 The European Council of Ministers has been renamed the 'Council of the European Union' by the Maastricht Treaty. It is made up of representatives (normally ministers) of the Member States and is the only EC body with plenary power to pass EC secondary legislation – although it can delegate legislative powers to the Commission. The Council's broad duties include ensuring the co-ordination of the general economic policies of the Member States and empowering the Commission to implement the Council's decisions.

4.17 The Council can only enact legislation proposed by the Commission and must observe complex procedures which involve the European Parliament. The Parliament was, under the Treaty of Rome, simply an advisory body, but this role has changed as a result of the Single European

Act and the Maastricht Treaty, in that it can now be said to share legislative power with the Council. The Maastricht Treaty gives the Parliament new powers of 'co-decision' with the Council in certain areas. This empowers the Parliament to reject a Commission proposal, by an overall majority of its members, if agreement cannot be reached between the Parliament and the Council in a Joint Conciliation Committee. Effectively, new powers will give the Parliament the final veto over particular EC legislation. The Parliament may also propose amendments to draft legislation and request that the Commission produce new draft legislation.

4.18 The Council's original voting system allowed any single Member State to veto a proposal before the Council; however, the Single European Act introduced 'qualified majority voting' (QMV) in a number of areas to realise the Single European Market on time and QMV has now been extended to other areas (notably certain environmental legislation) by the Maastricht Treaty. The QMV system is based on determination by a simple majority, where votes are weighed roughly in accordance with each Member State's population size.

The European Court of Justice

4.19 The Court of Justice can be described as the final court of appeal in relation to the interpretation and implementation of all EC legislation.

Much has been said about the sovereignty of the UK Parliament; however, accession to the Treaty of Rome immediately transferred considerable power to the Court of Justice. Article 219 of the Treaty provides:

> 'Member States undertake not to submit a dispute concerning the interpretation or application of this Treaty to any method of settlement other than those provided for therein.'

Article 164 goes on to say that the Court of Justice

> 'shall ensure that in the interpretation and application of this Treaty the law is observed.'

These and other EC provisions have effect in the UK by virtue of s 2(1) of the European Communities Act 1972, and s 3(1) of that Act makes it clear that an interpretation of an EC provision by a UK court must be in accordance with the principles laid down by the Court of Justice.

4.20 In *Van Gend en Loos v Nederlandse Tarief Commissie* [1963] CMLR 105, the Court of Justice made it clear that it saw European law as having sovereignty over national legal systems stating

> 'the Community constitutes a new legal order in International Law for whose benefit the States have limited their sovereign rights, albeit within limited fields, and the subjects of which comprise not only the Member States but also their nationals.'

Jurisdiction

4.21 The jurisdiction of the court can be briefly stated as follows:

- ensuring that Member States comply with their EC treaty obligations;
- ensuring that the actions and failures to act of the Council and the Commission (and, as a result of the Maastricht Treaty, the European Central Bank) are legal;
- to give preliminary rulings on EC law at the request of a national court or tribunal or EC institution;
- to grant compensation for damage caused by EC institutions; and
- to act as a court of appeal from decisions of national courts.

The court is able to render judgment against corporations or individuals within Member States and, if it does so, the judgment must be enforced by the relevant state's appropriate institutions.

In 1989, a Court of First Instance was formed to deal with a restricted range of matters. Decisions of that court can be appealed to the Court of Justice.

The Committee of Permanent Representatives (COREPER)

4.22 COREPER has the function of a secretariat to the Council of Ministers and has been given formal status by the Maastricht Treaty. It co-ordinates the preparatory work of the Council, scrutinises proposed legislation, and acts as a means of communication between the Commission and the Member States, and between the States themselves. The Committee also has delegated power to discharge some of the Council's functions, subject to formal agreement by the Council.

The European Council

4.23 The Maastricht Treaty has also endorsed the status of a further institution, named, confusingly, the 'European Council'. It arose from summit meetings of EC Heads of State and its role was declared after the summit meeting of June 1983:

> '... when the European Council acts in matters relevant to the European Communities it does so as the Council in the sense of the Treaties. However, in practice, it tends to confine itself to discussions in principle and broad political declarations, trying to unblock the community process and projecting the community as a going concern.'

EC LEGISLATION

The 'primary' legislation

4.24 The primary legislation is in the form of treaty provisions together with their protocols and annexes, and they provide the framework and basis for the Community as a whole, laying down the powers of the EC institutions and the decision-making procedures. Some treaty provisions are 'directly effective' in the sense that they ought to be followed by national courts. Other obligations have to be given effect within the legal systems of the Member States. In the UK, this is achieved by domestic legislation: thus the UK's accession to the EC was ratified by the European Communities Act 1972 (along with accession to Euratom and the ECSC) and the UK's acceptance of the Single European Act 1986 and the Maastricht Treaty were ratified by amendments to the 1972 Act.

The 'secondary' legislation

4.25 The 'secondary' legislation is in the form of regulations, Directives and decisions. This legislation is 'secondary' in the sense that it is enacted pursuant to and within the framework provided by the treaty provisions. Such secondary legislation may not be legally valid if it is in conflict with the treaty provisions, and challenges in the Court of Justice are often mounted on that basis. A national court does not have the power to pronounce any form of EC legislation to be legally invalid; however, it can refer the challenge to the Court of Justice for a preliminary ruling.

Regulations

4.26 Regulations are Community-wide legislative instruments. Article 189 of the Treaty of Rome states, inter alia, that:

> 'A regulation shall have general application. It shall be binding in its entirety and directly applicable in all Member States.'

Member States thus have no discretion as to whether regulations will or will not come into force and, when they do, they override any conflicting national legislation. Whilst this brings certainty to this method of legislation, it limits its range of uses, because many measures cannot be implemented in Member States until the necessary administrative infrastructure has been created. As such, regulations have tended to be used for technical matters only, such as prescribing standards. The progress of the Single Market has, however, created a change of emphasis away from flexibility of approach and regulations are becoming more wide-ranging.

Directives

4.27 Directives are addressed directly to Member States. They are designed to leave the choice of how they are implemented to the Member States and, normally, implementation is required within a specified period of time. Article 189 of the Treaty of Rome states (inter alia):

> 'A Directive shall be binding, as to the result to be achieved, upon each Member State to which it is addressed, but shall leave to the national authorities the choice of form and methods.'

Implementation normally takes place by the Member State changing its domestic law or administrative arrangements to comply with the Directive.

4.28 Some Directives are 'framework' Directives in that they establish only the broad frameworks of the regimes they address. These are then followed by 'daughter' Directives, which set out matters of detail. Other Directives are complete in themselves and directly address specific problems.

4.29 If a Directive is unconditional, sufficiently precise and the time-limit for its implementation has expired, then it may be held to be 'directly effective'. In other words, the Directive can be enforced in and through the national courts of individual Member States, although points of interpretation may be referred to the European Court of Justice by national courts. Section 2(1) of the European Communities Act 1972 provides that Community rules which are directly effective will come into force in the UK without further enactment.

'Emanations of the state'

4.30 Directives are enforceable only against bodies which comprise 'emanations of the state'. This requirement is normally construed as limited to governmental and quasi-governmental organisations; however, on 25 July 1994, the High Court held that privatised water companies could be 'emanations of the state' and directly subject to EC Directives: *Griffin and Others v South West Water Supplies Ltd* [1994], see ENDS Report 236, pp 38–9. There the High Court applied tests formulated by the European Court of Justice in ascertaining whether or not an organisation is an emanation of the state, namely:

(a) it must be providing a public service;
(b) it must be under the control of the state; and
(c) it must for that purpose possess special legal powers beyond those normally applicable on general laws between individuals.

It was held that South West Water satisfied all three criteria. As to (b), the control test, it did not matter that the state did not exercise day-to-day control over South West Water: the critical question was not whether the particular body was under that control, but whether the public service was.

Failure to implement

4.31 A citizen may need to assert the direct effect of a Directive where his Member State has failed to implement a relevant provision of the Directive and he is relying on it either to claim a right or as a defence to a charge under national law. A Directive may override inconsistent national law (see *R v Secretary of State for Transport ex parte Factortame* [1991] 1 AC 603) and national law has to be construed with regard to a relevant Directive even if it has not been implemented: *Marleasing SA v La Comercial Internacional de Alimentacion SA* (1992) 1 CMLR 305.

4.32 *In Frankovitch v Italian Republic* [1993] 2 CMLR 66, the European Court of Justice stated that, in principle, Member States would be required to compensate individuals for their failure to implement EC Directives:

> 'The full effectiveness of Community rules might be called into question and the protection of the rights which they conferred would be weakened if individuals could not obtain compensation where their rights were infringed by a breach of Community law for which a Member State was responsible.'

The court then stated the conditions for such liability. First, the Directive must involve rights conferred upon individuals and the content of those rights must be identified on the basis of the provisions of the Directive. Next, a link had to be established between the failure by the Member State to implement the Directive and the damage actually suffered by the person affected. In the absence of any Community rules on the subject, it was for the internal legal order of each Member State to determine the competent courts and appropriate procedural steps for obtaining compensation.

EC ENVIRONMENTAL POLICY

4.33 Article 2 of the Treaty of Rome states:

> 'The Community shall have as its task, by establishing a common market and progressively approximating the economic policies of Member States, to promote throughout the Community a harmonious development of economic activity, a continuous and balanced expansion, an increase in stability, an accelerated raising in the standard of living and closer relations between States belonging to it.'

It soon became clear, however, that economic expansion had to have regard to environmental constraints. In 1972, the Heads of State of the Member States requested the Commission to draw-up an EC environmental policy. This followed a declaration that 'economic expansion is not an end itself'. The Commission responded by producing the First Action Programme on the Environment. This has been followed by Action Programmes in 1977, 1982, 1987 and 1992. The constitutional basis for these Action Programmes is somewhat unclear; none the less, they have considerable persuasive and moral effects among the Member States and give a strong indication of the way in which EC environmental policy is going to develop.

The Fifth Action Programme

4.34 The Fifth Action Programme on the Environment is entitled 'Towards Sustainability. A European Community Programme of Policy and Action in Relation to the Environment and Sustainable Development'. The Programme will guide the EC's environmental policy until the year 2000, although it will be reviewed in 1995. It places great emphasis on the need to integrate environmental protection into policies at every level of government. High priority is given to promoting changes in patterns of behaviour and closer involvement in environmental protection by public authorities, businesses and the general public. It states that these goals should be pursued in a 'spirit of shared responsibility'.

The Programme is different from its predecessors in that it sets out quantitative targets for such matters as air pollution and waste management.

Improved access to environmental information is stressed together with the need for eco-labelling. Of particular concern to businesses are the proposals for legal requirements for environmental information in companies' annual reports.

4.35 The Fifth Action Programme promised new EC legislation in the following areas:

- the improvement of information on the environment;
- the widening of environmental research and development;
- the development of informational and educational schemes; and
- the promotion of land and sector-specific planning.

4.36 Three new EC environmental bodies were proposed:

- an environmental advisory forum with representatives from consumer organisations, trade unions, non-governmental organisations and local and regional government;
- a network of government officials responsible for overseeing the putting into practice of environmental legislation; and
- an environmental policy implementation group made up of both Commission and Member-State government officials.

EC ENVIRONMENTAL LEGISLATION

4.37 Until the Single European Act 1986 came into force, there was very little in the Treaty of Rome which could be construed as allowing the EC to create environmental legislation. In practice, however, Arts 100 and 235 of the Treaty were pressed into service.

Article 100 deals with the harmonisation of laws within Member States in order to ensure that businesses within the EC compete on level terms. In *Commission v Italy* [1981] 30 CMLR 331, the European Court of Justice stated:

> 'Provisions which are made necessary by considerations relating to the environment and health may be a burden on the undertakings to which they apply, and if there is no harmonisation of national provisions on the matter, competition may be appreciably distorted.'

Article 235 provides that the institutions of the EC can take appropriate measures to achieve any of the objectives of the Community that cannot be achieved through other powers. Directive 74/409, on the conservation of wild birds, was produced under Art 235: see **12.28**.

The Commission would often add robustness to environmental Directives by citing Arts 100 and 235 together. Thus, the environmental assessment Directive 85/337 was produced under both: see **10.22–10.25**.

Effects of the Single European Act 1986

4.38 The Single European Act 1986 (SEA) devoted a complete chapter to environmental matters, thereby obviating the question of whether or not these fell squarely within the constitutional competence of the EC's institutions. In particular, the SEA introduced Arts 100a, 130r, 130s and 130t.

4.39 The 1986 Act provides two different voting procedures for environmental legislation. Article 130s provides that the Council of Ministers must act unanimously in deciding what action is to be taken by the EC. Article 100a introduced QMV (see **4.18**) for proposals concerning health, safety, and environmental or consumer protection, and stated that the standards should be high.

4.40 Article 130r lists the objectives of the EC as including the preservation, protection and improvement of the quality of the environment, the protection of human health and the prudent and rational utilisation of national resources. Article 130r(2) provides a number of important environmental principles, including the preference of preventative action to remedial measures (see **4.44–4.45** below) and, importantly, the 'polluter pays' principle (ie that the polluter should pay for the costs of measures taken to protect the environment as a result of his activities). Article 130r(4) provides a principle of 'subsidiarity' in relation to environmental matters (see **4.46–4.49** below).

4.41 Article 130t provides that the measures adopted pursuant to Art 130s shall not prevent any Member State from maintaining or introducing more stringent protective measures (see **4.49**).

'Sustainable development'

4.42 In December 1988, the EC Heads of State issued a Declaration on the Environment that 'sustainable development must be one of the over-riding objectives of all EC policies'. The phrase 'sustainable development' is somewhat imprecise and has led to considerable speculation as to exactly what it is intended to achieve.

In the UK White Paper *This Common Inheritance* (published September 1990); the UK Government set out the broad principles of Britain's environmental strategy for the future. Paragraph 1.14 states:

> 'The starting point for this Government is the ethical imperative of stewardship which must underline all environmental policies. Mankind has always been capable of great good and great evil. This is certainly true of our role as custodians of our planet. The Government's approach begins with the recognition that it is mankind's duty to look after our world prudently and conscientiously. It was the Prime Minister who reminded us that we do not hold a freehold on our world, but only a full repairing lease. We have a moral duty to look after our planet and to hand it on in good order to future generations. That is what experts mean when they talk of "sustainable development"; not sacrificing tomorrow's prospects for a largely illusory gain today.'

The Maastricht Treaty has now amended Art 2 of the Treaty of Rome to provide that, among other things, the EC shall have as its task the duty to promote through the Community:

> '... a harmonious and balanced development of economic activities, *sustainable* and non-inflationary growth respecting the environment ...' [emphasis added]

The 'high level of protection' doctrine

4.43 The Maastricht Treaty created a new Art 130r(2), which provides that EC policy on the environment shall aim at

> '... a high level of protection taking into account the diversity of situations in the various regions of the Community.'

The new provisions expand an environmental policy first included in Art 100a(3) (as inserted by the SEA) to *all* aspects of EC environmental policy. The qualifying words 'the diversity of situations in the various regions of the Community' have been seen by some commentators as being a let-out clause which could be used as a justification for proposing or adopting measures that are pitched below the higher standards in the Community.

The precautionary principle

4.44 The Maastricht Treaty has inserted a new Art 130r into the Treaty of Rome, which states that EC policy on the environment 'shall be based on the precautionary principle', thus adding to the SEA principle that 'preventative action should be taken and that environmental damage should as a priority be rectified at source'. This precautionary principle is redolent of para 1.18 of *This Common Inheritance*:

> 'That does not mean that we must sit back until we have 100% evidence about everything. Where the state of our planet is at stake, the risks can be so high and the cost of corrective action so great, *that prevention is better than cure.* We must analyse the possible benefits and costs both of action and inaction. Where there are significant risks of damage to the environment, the Government will be prepared to take *precautionary action* to limit the use of potentially dangerous materials or the spread of potentially dangerous pollutants, even where scientific knowledge is not conclusive, if the balance of likely costs and benefits justifies it. This precautionary principle applies particularly where there are good grounds for judging either that an action taken promptly at comparatively low costs may avoid more costly damage later, or that irreversible effects may follow if action is delayed.' (Emphasis added.)

4.45 The practical effect of the doctrine is unclear. In *R v Secretary of State for Trade and Industry ex parte Duddridge* (1994) *The Independent*, 4 October, it was argued that the Secretary of State should apply the precautionary principle in siting power lines, because there was a possibility that the lines could pose a health risk. The High Court rejected that argument, thereby asserting that the precautionary principle is not a directly applicable legal obligation: see Tromans, 'High Talk and Low Cunning: Putting Environmental Principles into Practice' [1995] JPL 779.

'SUBSIDIARITY'

4.46 The concept of 'subsidiarity' is the EC's purported antidote to the allegation that it interferes unnecessarily in the everyday lives of those living in Member States. In broad terms, the concept is that the EC will interfere less in the internal affairs of Member States, because any decision not affecting other Member States should be taken at the lowest possible level, either nationally, regionally or locally. One of the fundamental problems with subsidiarity, however, is that it appears to mean exactly what different people want it to mean. Some see it as giving greater autonomy to Member States, whereas some see it as a meaningless concept which ought to give way whenever it is expedient.

A level playing field?

4.47 The concept of subsidiarity is of considerable importance in relation to environmental matters, because it is worth asking whether it is really necessary to have uniform environmental standards across the entire EC. Community-wide standards have the advantage of administrative neatness but are they justified on environmental grounds?

The problem is that climates and land conditions vary across the EC. For example, the UK is an island nation which enjoys a more than healthy amount of rain and, in many cases, the environmental impact of pollution is often less than it would be in other Member States. It is also necessary to question the purported axiom that the pursuit of environmental standards must be an international one because 'pollution does not respect national frontiers'. That assertion is far from a self-evident truth and depends upon consideration of particular types of pollution in particular locations. Pollution is very often only of concern to those within the country where it occurs. It would seem, therefore, that the desire of many commentators to see common environmental standards applied through the EC has as much to do with the level playing field concept as it does with an environmental justification.

The Danish bottles case

4.48 It is important to note that the subsidiarity principle is more evolutionary than revolutionary in character. EC law has always given Member States leeway to impose their own domestic environmental requirements – but within limits. In *Commission of the European Communities v Denmark* [1989] 54 CMLR 619 the European Commission challenged a Danish law which required beer and soft-drink containers to be returnable. The basis of the challenge was that the law was a form of discrimination against foreign manufacturers. This was alleged to be a breach of Art 30 (which promotes free trade within the Community); however, Art 36 permits national laws which are effectively restrictions on imports, provided that there is a genuine need to protect, among other things, the environment.

The European Court of Justice upheld the Danish laws, stating that it is permissible to use environmental protection as justification for such discrimination. Such a derogation from a free market should, however, be proportionate to the end to be achieved. The returnability requirement was therefore upheld on the basis of being a proper environmental restriction. A further requirement that only a limited number of container shapes was permitted was held to be disproportionate and therefore in breach of EC law.

The Danish bottles case is important because it makes it clear that individual Member States have the ability to pass environmental legislation that is stricter than in other Member States, thus interfering with free trade with the EC, provided that the motive is properly related to environmental protection and that the measures are proportionate to the end to be achieved.

Article 130t

4.49 Article 130t of the Treaty of Rome now provides that a Member State may, where legislation has been adopted under the environmental provisions of the Treaty, adopt more stringent protective measures. However, such measures must be notified to the Commission and must be compatible with the Treaty. This condition suggests that Art 130t cannot be used to disguise trade restrictions and, presumably, the proportionality test in the Danish bottles case would be relevant here.

Chapter 5

PRIVATE LAW LIABILITIES AND REMEDIES

INTRODUCTION

5.1 Much of this book deals with the regulation of contaminated land by statutory bodies, but it is important to recognise that the legal remedies available to ordinary citizens and organisations have a very important role to play and can involve not only the original polluter but also later occupiers in liability. This chapter will discuss these private law constraints primarily by reference to the common law. However, reference will also be made to certain private law remedies which derive from statute.

THE COMMON LAW

5.2 The majority of private actions take place by way of the common law. The common law is non-statutory in the sense that it is the system of law which has evolved in England and Wales, over the centuries, by way of the judgments of High Court judges. These judgments are reported in a variety of law reports and form a body of law because they will be followed or taken into account in later cases according to the 'doctrine of precedent'.

5.3 By its nature, the common law is an incomplete system. A principle of law can emerge only as a result of a contested case and there is no guarantee that crucial points of principle will ever be brought forward and contested. That uncertainty is compounded by the fact that precedents set decades (perhaps centuries) ago may remain unchallenged and binding on modern courts, despite the fact that these precedents may be completely out of step with modern society and its expectations.

The law of tort

5.4 This chapter is concerned with that particular branch of common law known as the 'law of tort', which deals with the liabilities that occur between parties otherwise than by formal arrangements such as contracts, leases and so on. Within the law of tort, it is necessary to examine the concepts of private nuisance, the so-called 'rule in *Rylands v Fletcher*', negligence and trespass.

Private nuisance and *Rylands v Fletcher* have been explored in depth by the House of Lords in the *Cambridge Water Co* case, which will be discussed below. The following subheadings dealing with private nuisance and *Rylands v Fletcher* will therefore cover matters of principle only.

THE LAW OF PRIVATE NUISANCE

5.5 Private nuisance has been defined as 'the unlawful interference with a person's use or enjoyment of land or some right over or in connection with it': *Read v Lyons* [1945] AC 156. Nuisance can arise by pollution of any of the environmental media, including running surface waters and percolating underground waters: see *Pride of Derby and Derbyshire Angling Association v British Celanese Ltd* [1953] Ch 149 and *Ballard v Tomlinson* (1885) 29 Ch 115. It has included unpleasant smells from a pig farm (*Bone v Seale* [1975] 1 WLR 797); and noxious fumes from a factory (*St Helens Smelting Company v Tipping* (1865) 11 HLC 642).

5.6 Whilst nuisance has been held to cover both damage to property and personal discomfort to its occupier, a person bringing a nuisance action can do so only if he has an interest in the affected land: *Malone v Laskey* [1907] 2 KB 141. If the complaint is one of pollution of a watercourse, then the owners or occupiers of the land in contact with the waterway will have rights of action. If it is percolating underground water, then any owner or occupier who can abstract water from that source may sue. Thus, whilst nuisance is often characterised as a 'neighbour' problem, it can clearly go way beyond adjoining land ownership. The nuisance arising from water pollution may easily impact on numerous properties some many miles away from the source.

5.7 Furthermore, a plaintiff is only entitled to relief if the interference with his land or his enjoyment of it results from the defendant's unreasonable use of his property: *Andreae v Selfridge & Company* [1938] Ch 1. In assessing what is reasonable, the courts have to strike a balance between the needs of neighbouring landowners. In *Harrison v Southwark and Vauxhall Water Company* [1891] 2 Ch 409, Vaughan Williams J stated:

> '... a man who pulls down his house for the purpose of building a new one no doubt causes considerable inconvenience to his next-door neighbours during the process of demolition; but he is not responsible as for a nuisance if he uses all reasonable skill and care to avoid annoyance to his neighbour by the works of demolition.'

A malicious intent (eg to annoy a neighbour) may turn a reasonable and non-actionable activity into a nuisance: *Hollywood Silver Fox Farm v Emmett* [1936] 2 KB 468.

5.8 Private nuisance is subject to a number of limitations:

- the locality test;
- the 'undue sensitivity' limitation;
- the defence of prescription; and
- statutory authority.

The locality test

5.9 A distinction is drawn between conduct causing damage to property and conduct causing personal discomfort only: *St Helens Smelting Company v Tipping* (1865) 11 ER 1483. In the latter situation it is necessary to have regard to the nature of the locality. In *Sturges v Bridgman* (1879) 11 Ch D 852, Thesiger LJ stated pithily that :

> 'What would be a nuisance in Belgrave Square would not necessarily be so in Bermondsey.'

There are, however, limits to this locality test. In *Rushmer v Polsue and Alfieri* [1906] Ch 234, Cozens-Hardy LJ stated:

> 'It does not follow that because I live, say, in the manufacturing part of Sheffield I cannot complain if a steam-hammer is introduced next-door, and so worked as to render sleep at night almost impossible, although previously to its introduction my house was a reasonably comfortable abode, having regard to the local standard; and it would be no answer to say that the steam-hammer is of the most modern approved pattern and is reasonably worked.'

5.10 The locality test has been complicated by *Gillingham Borough Council v Medway (Chatham) Dock Company* [1992] 3 All ER 923, where Gillingham Borough Council had granted planning permission for the use of a disused dock yard as a commercial port. Unfortunately, the operation of the port caused noise and disturbance to the neighbourhood, with the result that Gillingham then sought injunctions in public nuisance restricting the use of the port by heavy goods vehicles. In the High Court, Buckley J refused to grant the injunctions on reasoning which appeared to involve a defence of statutory authority (see **5.15**) and the proposition that the grant of the planning permission changed the character of the neighbourhood in a way that accommodated the full-time operation of the port.

5.11 The approach in the *Gillingham* case was considered and narrowed by the Court of Appeal in *Wheeler v J.J. Saunders Ltd* [1995] 2 All ER 697: a nuisance action involving the intensification of pig farming at a site by the erection of two pig-weaning units following the grant of planning permission. The court rejected a defence based on the principle in the *Gillingham* case. Peter Gibson LJ queried whether the *Gillingham* case had been rightly decided, and indicated that it might be right to refuse an injunction where a major development altered the character of a neighbourhood with wide consequential effects, such as required a balancing of competing public and private interests before permission was granted. That was not the situation in *Wheeler*. Staughton LJ said that it would be a misuse of language to describe what had happened in the instant case as a change in the character of the neighbourhood.

The 'undue sensitivity' limitation

5.12 In *Robinson v Kilvert* (1889) 41 Ch D 88, the plaintiff stored some paper in a warehouse above rooms let to the defendant. The defendant heated his rooms and damaged the plaintiff's paper; however, the plaintiff's subsequent action in nuisance failed because the paper was of a very sensitive nature. Thus, somebody who is hypersensitive to, say, smells, will find that his hypersensitivity is discounted in assessing whether or not the smells amount to a nuisance. However, this does not mean that the hypersensitive plaintiff has no chance of a remedy.

5.13 The defendant's heating in *Robinson v Kilvert* would not have affected normal paper but, had it been capable of doing so, then the plaintiff would have had a proper cause of action notwithstanding the sensitivity of his paper. Thus there was an actionable nuisance in *McKinnon Industries v Walker* [1951] 3 DLR 577, where poisonous gases which would have killed robust plants killed some (allegedly hypersensitive) orchids.

The defence of prescription

5.14 The defence of prescription arises where the relevant action has been carried out lawfully, openly and continuously for 20 years without the permission of the affected neighbour. The defence is a narrow one: in *Sturges v Bridgman* (1879) 11 Ch D 852, a confectioner had operated his noisy business for over 20 years without any problems; however, a doctor then built some consulting rooms which were affected by the noise. The defence of prescription did not apply because the 20-year period started only when the consulting rooms were constructed. Similarly, in *Bliss v Hall* (1838) 4 Bing NC 183, it was held that it is no defence to say that the plaintiff has just moved to the area and should have been aware of the nature of his future neighbours before doing so.

Statutory authority

5.15 If a statutory body is discharging its duties pursuant to a statute, then where the statute clearly exempts the defendant from liability, that statutory authority is a defence to any action in nuisance. The statutory authority covers not only the action specifically referred to by the statute, but also all relevant consequential actions. In *Allen v Gulf Oil Refining Limited* [1981] 1 All ER 353, the House of Lords held that a private Act of Parliament which allowed the acquisition of land at Milford Haven for an oil refinery also permitted the smells and disturbance created by its operation.

Liability under nuisance

5.16 Liability attaches to the person causing the nuisance, regardless of whether or not he is, or was at the time, the owner of the offending land. If the action is for damages only in respect of a concluded nuisance, then an erstwhile owner or occupier might be liable. Thus disposing of the land does not necessarily mean that liability has been avoided.

5.17 The fact that a nuisance arises from natural events or the action of a third party is not necessarily a defence. In *Goldman v Hargrave* [1967] 1 AC 645, a landowner was held liable when he failed to extinguish fully a fire which had been started by lightning. In *Leakey v National Trust* [1980] QB 485, a landowner was held liable for failing to prevent the natural slippage of soil onto neighbouring land.

If a nuisance is created by a trespasser, or arose before the current owner bought the offending site, then the current owner or occupier becomes liable if he adopts the trespassers' actions for his purposes or otherwise fails to abate the nuisance when he knew or ought to have known, by the exercise of reasonable care, of its existence: see *Sedliegh-Denfield v O'Callaghan* [1940] AC 880 and *St Anne's Well Brewery Company v Roberts* (1929) 140 LT 1. It is also worth noting that the occupier is not absolved from liability because the nuisance was caused by his independent contractor: see *Spicer v Smee* [1946] 1 All ER 489.

5.18 These decisions underpin one of the fundamental principles in buying land, namely that the prospective owner of a site should undertake some form of environmental assessment before entering into any binding commitment: see Chapter 13.

THE RULE IN *RYLANDS v FLETCHER*

5.19 In *Rylands v Fletcher* (1865) 3 H&C 774, the defendant's artificial reservoir flooded a nearby mine. The House of Lords held him liable to the mine owner on the basis that where a person, for his own purposes, brings onto his land and collects and keeps there anything likely to do mischief if it escapes, then he must keep it in at his peril and, if he does not do so, is prima facie liable for all the damage which is the natural and anticipated consequence of its escape. Liability under this rule can include not only the owner of the offending site but also those who have control or occupation of it. This includes lessees and licensees and, arguably, mortgagees in possession.

This principle has been construed by numerous later courts as being based on a form of strict liability; however, the House of Lords has now held that the key element involves a test of foreseeability: see discussion of *Cambridge Water Co* case below.

5.20 The rule does not apply:

- to things naturally on the land;
- where the plaintiff consented to the presence of the thing on the defendant's land or caused its escape;
- where the escape is due to an act of God or action by a stranger; or
- where the defence of statutory authority is upheld (see **5.15**).

The requirement that the defendant must have brought, collected and kept the non-natural thing on his land should mean that later owners or occupiers of the land would not be liable under *Rylands v Fletcher* for that deposit – although they might become liable in nuisance or for breach of statutory duty.

Rylands v Fletcher: non-natural use of land

5.21 From the outset, liability under *Rylands v Fletcher* has only arisen in relation to what the House of Lords then called the 'non-natural' use of the land – in that case the artificial accumulation of water. However, the phrase 'non-natural use' has been modified over time to the point where its applied meaning bears little relationship to its literal meaning. Uses have been held to be 'natural' where they are for the general benefit of the community. This requirement has created considerable uncertainty.

In *Rickards v Lothian* [1913] AC 263, for example, it was stated that liability would arise only where the use of the land involved 'some special use bringing with it increased danger to others and must not merely be the ordinary use of the land or such a use as is proper for the general benefit of the community': per Lord Moulton at p 280. In *Read v Lyons* [1947] AC 156, the House of Lords was not prepared to apply the rule to the storage of high explosives. Likewise, in *British Celanese Ltd v A.H. Hunt (Capacitors) Ltd* [1969] 2 All ER 1252, the manufacture of electrical components and storage of metal foil was held not to be a non-natural use of land, because the use was for the general benefit of the community.

Notwithstanding the uncertainties, the Canadian courts have held that methane produced from a landfill site fell within the rule, albeit on the basis that the use was said to be 'selfish and self-seeking': *Gertsen v Municipality of Metropolitan Toronto* (1973) 41 DLR (3rd) 646. This particular aspect of the rule has been modified by the *Cambridge Water Co* case (see below).

CAMBRIDGE WATER CO LTD v EASTERN COUNTIES LEATHER PLC

5.22 Fame is either welcome or unwelcome, and in the case of Eastern Counties Leather plc (ECL), the publicity it received in the early 1990s was of the unwelcome variety, because ECL became the focal point of a high-level debate on liability for pollution. ECL had, since 1879, pursued its business as a manufacturer of leather at Sawston, Cambridgeshire. The tanning process involved degreasing pelts and ECL had used solvents to do so since the early 1950s, including perchloroethene (PCE). Up until 1971 there were regular spillages of solvent onto the concrete floor of the tannery. At that time ECL were unaware that the spillages could cause an environmental problem.

However, a problem *was* caused to the Cambridge Water Co (CWC). CWC was responsible for supplying drinking water to a population of about 275,000, including the city of Cambridge. In 1979 CWC opened a borehole outside Sawston, which represented 12.5% of CWC's total resources. At that time the water met all drinking-water standards, but in 1980 an EC Directive was issued which, with effect from 1985, specified the acceptable levels of PCE in drinking water. CWC found that the water from its Sawston borehole did not meet that standard and, since there was no economic way of removing the PCE, the borehole had to be taken out of operation and an alternative source of supply found. CWC lost almost £1m as a result and sued those they believed had originated the PCE pollution – including ECL. It appeared that the PCE spillages at ECL's premises had seeped through the concrete floor and found their way into the underground waters which supplied the CWC borehole some 1.3 miles away.

5.23 In the event, only the action against ECL proceeded. It was based on negligence, nuisance and the rule in *Rylands v Fletcher*. The Court of Appeal found ECL strictly liable for the contaminants on the basis of a nineteenth-century precedent, *Ballard v Tomlinson* (1885) 29 Ch D 115, thereby effectively creating the spectre of unlimited retrospective liability for landowners generally.

5.24 The House of Lords overruled the Court of Appeal decision and, in doing so, revisited nuisance and *Rylands v Fletcher*: see [1994] 1 All ER 53. As to nuisance, Lord Goff confirmed at p 71 that:

> '... if the user is reasonable, the defendant will not be liable for consequent harm to his neighbour's enjoyment of his land; but if the user is not reasonable, the defendant will be liable, even though he may have exercised reasonable [care and skill] to avoid it.'

Thus, there is an element of flexibility in terms of the 'reasonable use' test. Lord Goff went on to confirm that, if the plaintiff shows that the defendant acted unreasonably, he still has to demonstrate the damage was foreseeable.

5.25 Turning to the rule in *Rylands v Fletcher*, the House of Lords rejected the notion that it did not include an element of foreseeability. Liability was strict in the sense that the defendant may be held liable notwithstanding that he has exercised all due care to prevent the escape from occurring, but the plaintiff had to prove, as with nuisance, that the damage was foreseeable. In effect, the House of Lords tended towards the view that the rule in *Rylands v Fletcher* is a branch of nuisance dealing with isolated escapes. It remains to be seen how the courts will test foreseeability in this context, and whether it would be based on what damage an expert would have foreseen or merely the foresight of the ordinary land user: see *West v Bristol Tramways Company* [1908] 2 KB 14.

The House of Lords also provided clarification of the 'non-natural use' test in *Rylands v Fletcher*. The Court of Appeal had held that ECL's use of its land was a natural use; however, the House of Lords disagreed. Lord Goff stated at p 79:

> '... indeed I feel bound to say that the storage of substantial quantities of chemicals on industrial premises should be regarded as an almost classic case of non-natural use.'

He added that now it was recognised that foreseeability of harm is a pre-requisite of liability under the rule, the courts may feel less pressure to extend the concept of natural use to circumstances such as those in that case.

5.26 On the facts, the House of Lords found that ECL could not, in the 1970s, have foreseen that the spillages would infect the borehole in the way they did. Thus, nuisance and *Rylands v Fletcher* would have been inapplicable. The House of Lords examined *Ballard v Tomlinson* and came to the view that it had not, as the Court of Appeal believed, created a new principle of strict liability, but was based on either nuisance or *Rylands v Fletcher* and, as such, involved the same foreseeability test.

The implications of the *Cambridge Water Co* case

5.27 No doubt landowners in general breathed sighs of relief at the House of Lords decision, but one has to question whether that relief is soundly based. ECL were not liable, because, at the time of the spillages, they could not have foreseen either that the spillages would have contaminated an aquifer or that such contamination would have affected a borehole some 1.3 miles away. However, the fact that the *Cambridge Water Co* case came to court at all has, by itself, altered our state of knowledge on the subject. It must, by now, be apparent to all professionally concerned in property that on-site spillages may have widespread off-site implications. That, by itself, alters the base-line for foreseeability. Furthermore, science has progressed considerably since the 1970s. Indeed, many of the current techniques for studying the flow characteristics of pollutants in this situation were created by the experts employed by CWC in tracing the sources of the PCE!

5.28 The facts of the *Cambridge Water Co* case are unique and have to be viewed with considerable care. Had the spillages occurred in 1980, then the House of Lords might have taken a different view on foreseeability, and similar spillages occurring in the late 1990s could easily result in liability. It is also worth remembering that many underground deposits remain where they are unless disturbed by works on the surface. One could, therefore, have a situation where the original polluter is not to blame for the original deposit, but a later occupier becomes liable for any damage which is the foreseeable consequence of its movement. It is opined that the actions of that later occupier will be judged by the standards and knowledge prevailing during his occupation, including the fact that environmental assessments were readily available to him.

Finally, one has to remember that certain statutory liabilities appear to have retrospective effect in any event, including s 161 of the Water Resources Act 1991: see **7.15–7.18**.

5.29 The less publicised sequel to the House of Lords action was that ECL had to come to terms with the National Rivers Authority (NRA) over remedial action. It appears that the NRA believed that a solution involving 10–20 pumping boreholes and costs of £2m was the best solution, but eventually settled for a solution costing £50,000 plus the costs of pumping contaminated groundwater: see ENDS Report 233.

NEGLIGENCE

5.30 Negligence is a fault-based tort, in the sense that the plaintiff must prove that the defendant was not only the actual cause of the relevant damage but also that he was blameworthy. The plaintiff must show that:

- the defendant owed him a 'duty of care';
- the defendant was in breach of that duty; and
- the plaintiff has suffered damage as a result of that breach of duty.

The classic description of the duty of care comes from *Donoghue v Stevenson* [1932] AC 562, where Lord Atkin stated 'you must take reasonable care to avoid acts or omissions which you can reasonably foresee would be likely to injure your neighbour'. Neighbours were 'persons who are so closely and directly affected by my act that I ought reasonably to have them in contemplation'.

A defendant will be in breach of the duty of care if his conduct fell below that which a person of reasonable skill and judgement should have done in the particular circumstances.

Finally, the defendant is liable only for damage which a reasonable man would have foreseen in the circumstances. Thus, reverting to the facts of the *Cambridge Water Co* case, a spillage of solvent in the 1970s would not, at that time, have been reasonably foreseen as causing damage to a drinking-water borehole some miles away and, as such, a negligence claim in those circumstances would (even if it passed all other tests for negligence) fail to pass this test.

5.31 Negligence has an advantage over trespass and nuisance for the plaintiff in that he does not need an interest in any neighbouring land – an injury to his person or movable property is actionable. Nor is liability necessarily restricted to the polluter. It may impact on persons who were under a duty to warn the person suffering the injury of his peril. In *Scott-Whitehead v National Coal Board* (1987) 53 P&CR 263, a regional water authority was held liable for failing to warn a farmer that the water he was abstracting was polluted by chlorine, with the result that his crops were damaged. That particular decision may need to be revisited as a result of *Murphy v Brentwood District Council* [1990] 2 All ER 908 (where the House of Lords redefined the liabilities of public authorities in this type of situation), but it serves to illustrate the broad principle.

TRESPASS

5.32 Trespass may arise from a direct interference with another person or his property without lawful excuse where that interference is intentional or negligent. Trespass is actionable without proof of fault on the part of the defendant. Placing a substance on another person's land without permission or excuse can be trespass, as can causing somebody to inhale an unpleasant gas.

5.33 Trespass appears to be a remedy of wide application, but it is severely restricted by the requirement of directness. In *Esso v Southport Corporation* [1956] AC 218, a tanker discharged oil at sea which was carried onto a beach at Southport. In the House of Lords, Lords Tucker and Radcliffe stated that the pollution to the beach was not trespass, because it was not inevitable that the action of wind, waves and tide would have brought the oil to the beach. Those opinions can be usefully compared with *Jones v Llanrwst Urban District Council* [1908–10] All ER Rep 922, where untreated sewage polluted a river and collected on banks owned by the plaintiff. An action in trespass was successful, because it was inevitable that the sewage would be carried downstream to the plaintiff's bank.

The plaintiff's burden of proof is a difficult one and he would be in severe difficulty where the pollution is borne by air or by way of underground waters.

BREACH OF STATUTORY DUTY

5.34 Not all private law actions arise from the common law: some arise from statute as an adjunct to statutory regulatory regimes and can be powerful weapons in the plaintiff's armoury. The Environmental Protection Act 1990 (EPA) has a useful supplemental provision in relation to 'controlled waste' (see **8.3** for a definition), where any person who deposited controlled waste or knowingly caused or knowingly permitted it to be deposited so as to cause an offence under ss 33(1) or 63(2) of the Act is liable for damage caused by the deposit: s 73(6). There is nothing to indicate that there needs to be a criminal conviction under ss 33(1) or 63(2) as a prerequisite to the civil action and the section appears to create strict liability in relation to damage caused, because there is no reference to foreseeability or the like.

Neither the plaintiff nor defendant under s 73 has to have any interest in land. The defendant is personally liable by virtue of his breach of the relevant EPA provisions and the plaintiff may claim in respect of damage to his person or real or movable property. Nor is there anything which confines any action to pollution damage – damage caused by the movement or presence of inert material appears to fall within the provision.

The injured party cannot rely on s 73(6) if he voluntarily accepted the risk of the damage being caused or it was wholly due to his fault. The matters which may be proved as a defence under s 33(7) of the EPA may also be proved as a defence to a civil action under s 73(6).

If the cause of the complaint amounts to a statutory nuisance, then s 82 of the EPA allows any 'person aggrieved' to apply to a magistrates' court for an abatement notice and that can be coupled with a claim for compensation under the Powers of the Criminal Courts Act 1973: see Chapter 9.

PUBLIC NUISANCE

5.35 Public nuisance contains many elements similar to private nuisance, but the relevant state of affairs must materially affect the reasonable comfort and convenience of life of a class of the public in general as opposed to a single individual. It is a question of fact in each case as to whether the number of people affected by a nuisance is enough to make it a public nuisance. In *R v Lloyd* [1802] 4 Esp 200, it was held that the occupiers of three chambers in Clifford's Inn were not a wide-enough class.

A public nuisance is a crime actionable through the Attorney-General only, but local authorities may seek injunctions under s 222 of the Local Government Act 1972. A private person may take independent civil action, but only if he has suffered special damage beyond that suffered by the affected class as a whole: *Attorney-General v Basingstoke* (1876) 45 LJ Ch 726.

OCCUPIERS' LIABILITY

5.36 Most of this chapter deals with the consequences of pollution emitted from contaminated land; however, it is necessary to remember that on-site contamination may affect the health and safety of those visiting the site. The occupier owes visitors a 'common duty of care' under the Occupiers' Liability Act 1957 and breach of that duty may make him liable in damages to an injured visitor.

5.37 The 'common duty of care' is defined by s 2(2) of the 1957 Act as

> '... a duty to take such care as in all the circumstances of the case is reasonable to see that the visitor will be reasonably safe in using the premises for the purposes for which he is invited or permitted by the occupier to be there.'

Valuable guidance can be gained from the Health and Safety Executive's publication, *Protection of Workers and the General Public during the Development of Contaminated Land* (HMSO, 1991).

Visitors encompassed by the common duty of care include the usual range of invitees and licensees. The duty can be modified or excluded by agreement or other measures.

5.38 The Occupiers' Liability Act 1984 deals with trespassers and imposes liability on occupiers where:

(a) the occupier must have been aware of the danger or had reasonable grounds to believe that it existed;
(b) he knew or had reasonable grounds to know that the uninvited entrant either was or might come into the vicinity of the danger; and
(c) the risk of injury to the uninvited entrant resulting from that danger was one against which in all the circumstances of the case the occupier might reasonably be expected to offer the uninvited entrant some protection.

That duty to trespassers can be discharged by appropriately sited and worded notices: s 1(5).

The 1957 and 1984 Acts both impose liability on 'occupiers' of land. An occupier can be any person with a significant degree of control over the premises, including licensees and tenants: *Wheat v Lacon* [1966] AC 522. Arguably, a mortgagee in possession could be an occupier for these purposes.

THE PROFESSIONAL DUTY OF CARE

5.39 As noted in **14.26**, many insurers now exclude cover for gradual pollution incidents. This has led to concerns that developers and land owners will try to pass liability on to their professional advisers either by indemnities or by professional negligence actions. Advisers themselves ought to be aware of the signs of contamination and its possible consequences. For example, it is difficult to believe that an architect who notices that a large area on an open site is bare of any natural growth ought not to be put on guard that this could be due to contamination – and, arguably, he ought to advise his client to carry out an environmental assessment. Likewise (after the publicity given to the *Cambridge Water Co* case), how could any competent lawyer fail to notify his client that a proposed purchase site once housed a tannery?

5.40 The standard of care expected of a professional adviser at common law was summarised, in very broad terms, by Bingham LJ in *Eckersley v Binnie and Partners* [1988] 18 Con LR1:

> 'A professional man should command the corpus of knowledge which forms part of the professional equipment of the ordinary member of his profession. He should not lag behind other ordinarily assiduous and intelligent members of his profession in knowledge of new advances, discoveries and developments in his field. He should have such awareness as an ordinarily competent practitioner would have of the deficiencies in his knowledge and limitations on his skill. He should be alert to the hazards and risks inherent in any professional task he undertakes to the extent that other ordinarily competent members of the profession would be alert. He must bring to any professional task he undertakes no less expertise, skill and care than other ordinarily competent members of his profession would bring, but need bring no more. The standard is that of the reasonable average … it is the standard prevailing at the time of his acts or omissions which provide the relevant yardstick. He is not … to be judged by the wisdom of hindsight.'

5.41 Reliance upon the common law is never prudent where significant sums of money or substantial risks are involved. The wise developer or landowner will therefore ensure that the appointment of a professional adviser is, where appropriate, subject to a specific written duty to have regard to any contamination of the site and its potential for off-site pollution. Consideration needs to be given to whether such contractual liability should cover potential claims from third parties or increased project costs or both. Regardless of whether or not the expert's position is reduced to writing, his client ought to check that the expert has adequate insurance cover in respect of types of claims (ie whether liability in connection with contamination is excluded), the financial maximum covered and the period of insurance (ie a

12-month policy is of little use in a 5-year project). The reality is that the insurance industry has reacted to increased awareness of environmental liability by reducing the cover available (see Chapter 14).

LIMITATION PERIODS

5.42 Non-personal injury actions for damages (not injunctions) based on negligence, nuisance or the rule in *Rylands v Fletcher* may not be commenced more than 6 years after the damage accrues, and the relevant period for personal injury cases is 3 years: Limitation Act 1980. The question of time-limits has caused some concern in environmental law.

In the *Cambridge Water Co* case, Ian Kennedy J, at first instance, was extremely reluctant to make an award of damages against ECL on the basis of actions which, when committed 15 years earlier, were not actionable nuisances or negligence. The House of Lords did not agree. Lord Goff pointed out that neither tort comes into being until actual damage is caused to the plaintiff – regardless of when the pollutant was emitted into the environmental media. Thus, if an aquifer was contaminated in 1970 and a potential plaintiff suffers patent property damage in 1995, he has until the corresponding date in 2001 to start his action.

5.43 The situation is complicated by the fact that damage does not always manifest itself immediately. Hidden damage to the foundations of a building may remain undetected for decades. In this type of case, damage is deemed to accrue to the plaintiff only when he ought to have become aware of it, and the position is governed by the Latent Damage Act 1986. The 1986 Act provides that, in non-personal injury negligence cases, the usual 6-year limitation period is extended, where the damage is latent, so that a claim outside that period has to be made within 3 years of the date that the plaintiff (or any person who could have taken action before him) had both sufficient knowledge of the damage and the right to bring the action. However, this is subject to a long-stop period of 15 years. Thus, if latent damage was done in 1995, the long-stop date is in 2010.

These time-limits have no relevance to nuisance actions for injunctions where the nuisance has been continuing for a considerable period of time, subject to the defence of prescription: see **5.14**.

INJUNCTIONS

5.44 An injunction may be mandatory, requiring that certain actions be carried out, or prohibitory, requiring that a described state of affairs be stopped. A final injunction will be granted only after a full trial, but a court may grant an injunction pending the full trial (an interlocutory injunction) at an early stage if the balance of convenience is in favour of doing so: *American Cyanmid v Ethicon* [1975] AC 396. A court may grant a 'quia timet' injunction to prevent an anticipated tort before damage occurs.

DAMAGES

5.45 Damages in tort are calculated as being the sum of money which (so far as money can do so) will put the plaintiff back into the position he would have enjoyed but for the defendant's wrong-doing. If two defendants have both joined in a single tort giving rise to a single injury or set of injuries to the plaintiff, then they will be jointly and severally liable to the plaintiff, in that each will be responsible for all of the damages. If two defendants have caused different damage to the plaintiff (albeit at the same time), then each will be liable for only a proportion of the total damage: see Civil Liability (Contribution) Act 1978. If the defendant has acted cynically and calculated that the profit he will make will exceed the damages he would normally suffer, then the plaintiff might be entitled to exemplary damages: see *AB v South West Water Services* [1993] 1 All ER 609.

Chapter 6

CRIMINAL LIABILITY

INTRODUCTION

6.1 The purpose of this chapter is to explain the criminal law concepts and terminology which appear in some of the following chapters. Criminal liabilities within particular regimes (ie water pollution, statutory nuisance, etc) will be discussed in later chapters; however, there are themes and concepts which appear repeatedly in the legislation. As such, it is sensible to discuss these matters in some detail at the outset, together with outlining relevant criminal procedures. It is important to note that criminal law is not independent of civil law. Thus a criminal prosecution for pollution which causes damage to another person may simply become a preliminary to, and evidence in, a much larger civil claim (see Chapter 5).

ELEMENTS OF CRIMINAL LIABILITY

6.2 As a generality, every crime requires a mental element (known as the 'mens rea') and a physical element (known as the 'actus reus'). The mental element may be connoted by expressions such as 'with intent', 'recklessly', 'knowingly' or 'permitting'. In broad terms, criminal liability is imposed on a person where he intends to, or foresees that he may, by his actions bring about the constituents of a crime. In some circumstances, however, 'strict liability' is imposed – which means that criminal liability can be incurred without proof of any mental element in the accused and where there may be no negligence or default by him.

6.3 Criminal liabilities can arise under both common law and statute; however, in the main it is only necessary to consider statutory offences for the purposes of this book. It is possible to obtain the primary definition of a particular offence directly from the words of the relevant statute. For example, s 80(4) of the Environmental Protection Act 1990 (EPA) provides that if the recipient of an abatement notice in respect of statutory nuisance

> 'without reasonable excuse, contravenes or fails to comply with any requirement or prohibition imposed by the notice, he shall be guilty of an offence.'

In this example, the words 'without reasonable excuse' prescribe a 'statutory defence' in the sense that a person charged under s 80(4) may, if he believes he can prove it, argue that, whilst he contravened or failed to comply with the notice, he did so in circumstances which excuse criminal liability. Section 80(7) provides another statutory defence, namely where the defence proves that 'best practicable means' have been employed to prevent or to counteract the statutory nuisance.

6.4 The relevant statute will also prescribe the appropriate mode of trial (see **6.33**) and penalty. The majority of crimes punishable in magistrates' courts will involve fines calculated on the 'standard scale'. The five levels of fine on the standard scale are currently:

Level 5	–	£5,000
Level 4	–	£2,500
Level 3	–	£1,000
Level 2	–	£500
Level 1	–	£200

Some environmental offences carry a statutory maximum fine of £20,000 in the magistrates' court, eg the unlawful deposit of controlled waste contrary to s 33 of the EPA.

The burden of proof

6.5 The burden of proving that a criminal offence has been committed always falls squarely on the prosecution. The defendant is always innocent until proven guilty and it is up to the prosecution to prove its case beyond all reasonable doubt. However, there is one important exception to this principle, namely that the burden of proving a statutory defence falls on the defendant. He must prove his statutory defence 'on the balance of probabilities', in other words that his defence is more probable than not.

If the defendant believes he is entitled to argue 'reasonable excuse', then it is up to him to raise the defence and prove it – he cannot merely assert that he has a reasonable excuse and then sit back. If the defence is that he had a licence or other permission, then it is up to the defendant to prove its existence: see s 101 of the Magistrates' Courts Act 1980 and *R v Edwards* [1975] QB 27.

To 'cause' or 'knowingly permit'

6.6 The most straightforward criminal situation is where a person is caught red-handed in the act of committing a crime, for example the burglar who is caught making away with his haul. Unfortunately, things are not always that simple. The burglar may have an accomplice waiting in a car, in which case the burglar is charged with carrying out the burglary and his accomplice is charged with 'aiding and abetting'. Here we have two methods of involvement in the crime, but there are many more including 'causing' and 'knowingly permitting' crimes. One example is s 85 of the Water Resources Act 1991 (WRA), which provides the criminal offence of *causing* or *knowingly permitting* any poisonous, noxious or polluting matter or solid waste to enter controlled waters (see Chapter 7).

'Cause'

6.7 It is often said that a word derives its meaning from the context within which it is used, an aphorism which is particularly apt in attributing meaning to the word 'cause'. To give a familiar example from road traffic law, there are a number of offences which arise by 'using' a vehicle in particular circumstances or 'causing' or 'knowingly permitting' the vehicle to be so used. The three phrases are juxtaposed to create three different modes of involvement. For example, A may be found guilty of 'using' B's vehicle without insurance, whereas B may be found guilty of 'causing' or 'knowingly permitting' the offences if he allowed A to use his car. In this context, B would be found guilty despite the fact that he did not physically participate in driving the car; and, also in this context, the word 'cause' has a very particular meaning which involves a mental element (or 'mens rea') on the part of the defendant.

In the House of Lords case, *McLeod v Buchanan* [1940] 2 All ER 179, Lord Wright said:

> 'To "cause" the user involves some express or positive mandate from the person "causing" to the other person, or some authority from the former to the latter, arising in the circumstances of the case.'

In this context, therefore, causation is a situation created by the defendant and with his knowledge where the actual act (ie 'using') is executed by another person.

6.8 The situation is, however, radically different where the statute creates an offence only to 'cause' or 'permit' a given state of affairs, without including the word 'use'. In such a case, to 'cause' equates to 'to do' and a mental element (ie 'mens rea') does not need to be proved by the prosecution; in other words, the offence is one of 'strict liability'. As noted above, the water pollution offence under s 85 of the WRA refers to 'causing' or 'knowingly permitting' only; it does not juxtapose any third phrase connoting doing.

6.9 The effect of this strict liability formula can be seen in *Alphacell v Woodward* [1972] 2 All ER 475, where the defendant company had polluted the River Irwell when washing manilla fibres, a raw material for paper-making. It was prosecuted for causing polluting matter to enter the river under what is now s 85. The House of Lords held that the prosecution did not need to prove that the company had knowingly, intentionally or negligently caused the polluted water to enter the river: the company had caused the pollutant to enter the river by their positive and deliberate acts in building and operating the system which led to the overflow into the river, and that was sufficient for liability.

6.10 In *CPC (UK) v National Rivers Authority* (1994) 235 ENDS Report 41, CPC was convicted of causing water pollution in contravention of s 85 of the WRA. The discharge was caused by a break in CPC's piping system. It was accepted that the break was latent and occurred before CPC bought the site; apparently the fault was the result of bad workmanship by subcontractors previously employed by CPC's predecessors. The Court of

Appeal held that it was not necessary for the National Rivers Authority (NRA) to show that CPC was the one and only cause of the pollution: the offence was committed if the jury were satisfied that CPC had caused the escape, irrespective of the role played by its predecessors' subcontractors. In effect, CPC was guilty of 'causing' the pollution by operating its factory with a faulty pipe: see also *Attorney-General's Reference No 1 of 1994* [1995] 241 ENDS Report 42.

6.11 The CPC case has to be carefully compared with another water pollution case, *Wychavon District Council v National Rivers Authority* [1993] 2 All ER 440, where Wychavon operated sewers on behalf of Severn Trent Water. A sewer became blocked, apparently, by solid material from a nearby hospital and raw sewage entered the River Avon. There were no inherent defects in the sewer itself. The Divisional Court held that the NRA had been wrong to prosecute Wychavon for causing water pollution contrary to what is now s 85 of the WRA. Applying the principles in the *Alphacell* case, it was held that the council had not performed any positive or deliberate act which brought about the water pollution.

6.12 In *Wrothwell Ltd v Yorkshire Water Authority* [1984] Crim LR 43 the director of a company poured herbicide into a drain believing that it led to a sewer when, in fact, it led to a stream. He was found guilty of causing pollution to the stream, notwithstanding that he had not intended to pollute it.

6.13 One defence to this type of charge is for the defendant to demonstrate that the pollution was actually caused by a trespasser: *Impress (Worcester) Ltd v Rees* [1971] 2 All ER 357.

'Knowingly permitting'

6.14 The *Alphacell*, *Wychavon* and *Wrothwell* cases could have been decided differently if the charges had been based on 'knowingly permitting'. In the *Wychavon* case (see **6.11**), Watkins LJ indicated that there were facts in the evidence which could have amounted to 'knowingly permitting'.

6.15 In *Price v Cromack* [1975] 1 WLR 988, a farmer allowed a company to discharge animal wastes into a lagoon on his land. The lagoon leaked and polluted a nearby river. The farmer was charged with causing the pollution but was acquitted because he had only permitted the accumulation of waste. Arguably, the prosecutor would have succeeded had he chosen to charge the farmer with knowingly permitting – provided, of course, that the farmer knew of the leakage. In order to succeed in a prosecution based on 'knowingly permitting' a contravention, the prosecution must always prove knowledge on the part of the defendant.

6.16 It is often the case that the current owner or occupier of a site is not aware of waste buried there and that pollutants from the waste may be percolating through the subsoil for considerable distances. If such migrating pollutants contaminated nearby land or watercourses, then the owner or occupier would not, at that moment, be causing or knowingly permitting the discharge. But if he then became aware of the pollution, he would have to take such measures as are within his powers to stop it. Failure to do so could amount to knowingly permitting: see *Schulmans Incorporated Limited v National Rivers Authority* (1993) *Water Law*, January, p 25. Likewise, the defendant may be found guilty if he deliberately shut his eyes to the obvious or refrained from inquiring, because he suspected the truth but did not want to have his suspicions confirmed: see *Westminster City Council v Croyalgrange Ltd* [1986] 1 WLR 674. If pollution is caused by vandals, then the question of whether or not the owner or occupier knowingly permitted it may depend on whether or not the vandalism was foreseeable: *National Rivers Authority v Wright Engineering Co* [1994] 4 All ER 281.

Liability of landlord and tenant

6.17 The situation is more complicated where a landowner is not in day-to-day control because, for example, the land is held by a leaseholder. A recent case involving the Welsh Development Agency (WDA) demonstrates that a landlord is not necessarily liable for the actions of his tenants in this context. In *National Rivers Authority v Welsh Development Agency* (1992) JP 506 PC, the WDA set up an industrial estate which they leased to tenants. A tenant of one of the industrial units released pollutants into a stream and the NRA unsuccessfully prosecuted the WDA. The High Court stated that the crucial issue was whether or not the WDA's positive or deliberate act caused the discharge: Cooke J held that the question must be answered in the negative.

6.18 It is important that, before entering into a formal tenancy, both proposed landlord and tenant make themselves aware of the potential criminal liabilities which could arise from the site and then draft the lease accordingly: see Chapter 14. It would be unfortunate if the holder of a short lease suddenly finds that he has become liable to remedy the pollution caused by the freeholder or a previous leaseholder.

It is not always practical to insist on comprehensive site surveys before entering a short lease, so the leaseholder would be well advised to ensure that the freeholder or head-lessee, as the case may be, is placed under an ongoing obligation to remedy pollution caused otherwise than by the short leaseholder. On the other hand, somebody taking a long lease could, properly, be asked to shoulder certain specified responsibilities for remedial measures.

'Contravention'

6.19 In everyday parlance, one 'contravenes' something by a positive action only; however, the use of the term in statutes usually includes failure to act. For example, s 73 of the Control of Pollution Act 1974 (COPA) provides that 'contravention' includes a failure to comply with the provisions in question.

'Aiding, abetting, counselling or procuring'

6.20 The phrase 'aiding, abetting, counselling or procuring' describes a single offence, and a person who aids, abets, counsels or procures an offence is liable to the same penalties as for the principal offence itself. To 'aid' means to give help, support or assistance to a person: 'abet' means to invite, instigate or encourage; 'procure' means to produce by endeavour; and 'counsel' means to advise: see *Attorney-General's Reference (No 1 of 1975)* [1975] 2 All ER 684.

6.21 In *Johnson v Youden* [1950] 1 KB 544, Lord Goddard CJ stated:

> 'Before a person can be convicted of aiding and abetting – he must at least know the essential matters which constitute that offence. He need not actually know that an offence has been committed, because he may not know that the facts constitute an offence and ignorance of the law is not a defence.'

Criminal charges under environmental law often involve offences of 'causing' or 'knowingly permitting'. Two road traffic cases indicate that these terms are wide enough to cover being an accessory or aiding and abetting in any event and, as such, the defendant should more properly be charged with causing or knowingly permitting: see *Carmichael & Sons Ltd v Cottle* [1971] RTR 11 and *Crawford v Haughton* [1972] RTR 125.

A person can be found guilty of aiding and abetting only if he was present when the crime was committed: *R v Anthony* [1965] 1 All ER 440. If a person is involved at another time then the appropriate charge is that of 'accessory'.

A person may be found guilty of aiding and abetting if he is under a legal duty to act to prevent an offence and fails to do so: see *Crampton v Fish* (1969) 113 SJ 1003 and *Carter v Richardson* [1974] RTR 314.

Conspiracy

6.22 Section 1(1) of the Criminal Law Act 1977 makes it an offence for two or more people to agree to commit a criminal offence – regardless of whether or not that offence is committed or is capable of being committed.

The agreement for conspiracy is not necessarily something settled in one go. One can have a 'chain' conspiracy, where A enrols B into the plot who later enrols C and so on. One can also have a 'spoke' conspiracy, where one person is at the hub and enrols others separately – perhaps unknown to one another.

If the prosecution wishes to charge conspiracy to commit an offence in addition to a substantive charge on the offence itself, then the joinder has to be justified: see *Practice Direction* [1977] 2 All ER 540.

The penalty for conspiracy is prison for a period not exceeding the maximum provided for the offence the defendant conspired to commit (where the offence carries a prison sentence) and an unlimited fine.

CORPORATE LIABILITY

6.23 The law normally treats a company as a legal person which exists separately from its directors, shareholders and employees. A company can commit crimes as a company and, aside from not being put in gaol, can incur the same penalties as a human criminal.

A company has a 'brain', in its directors and managers, and 'hands', in its employees, but it differs from a human being in that brain and hands are not always in direct communication. Save for the smallest of them, companies are usually departmentalised and, frequently, involve geographical separation between headquarters and branch offices. These characteristics bear on the way in which liability attaches to companies – there are two routes. First, by way of direct liability, that is by showing that the company, as a person, committed the crime. Secondly, by vicarious liability, that is by showing that, whilst it is not directly liable, the company is liable for the acts of its employees.

Direct liability

6.24 Until very recently the law has been understood to be that a company would be directly liable for a crime only if the criminal act happened as a result of a decision by the company's 'controllers'. In *Tesco Supermarkets v Nattrass* [1972] AC 153, Lord Reid indicated that the 'controllers' include the board of directors, the managing director and perhaps other superior officers of the company who carry out the functions of management and speak and act as the company. In that case it was held that a branch manager of a supermarket was not sufficiently highly placed within the Tesco hierarchy to rank as a controller. Lord Pearson said:

> 'In the present case the company has some hundreds of retail shops, and it would be far from reasonable to say that every one of its shop managers is the same person as the company ... supervision of the details of operations is not normally a function of higher management: it is normally carried out by employees at the level of foremen, chargehands, overlookers, floor managers and "shop" managers (in the factory sense of "shop").'

In this context, where the law distinguishes between a company's 'brain' and 'hands', it identifies direct liability with the former only. The *Tesco* case has been considered recently in *Re Supply of Ready Mixed Concrete (No 2) Director General of Fair Trading v Pioneer Concrete (UK) Ltd and Another* [1995] 1 All ER 135, where the House of Lords held that, for the purposes of the Restrictive Practices Act 1976, a prohibition at the level of senior management against the making of an illegal agreement or arrangement which is ignored by the employees concerned does not amount to a defence. Such a defence may, however, succeed if the relevant statute merely requires that the employer 'take all reasonable steps' or 'take all such measures as are reasonable in the circumstances' (eg s 34 of the EPA) and he persuades the court that he did, indeed, do everything that was reasonable in the circumstances: see *Seaboard Offshore Ltd v Secretary of State for Transport* [1994] 1 WLR 541.

6.25 The *Tesco* case involved a prosecution under the Trades Descriptions Act 1968, where Tesco was raising a statutory defence which turned on the manager being 'another person' separate from Tesco. It also involved a non-environmental subject-matter. Both of these points now appear to have been relied upon by the High Court to narrow the effect of the *Tesco* case: *National Rivers Authority v Alfred McAlpine Homes East Ltd* (see **6.38**). Nor is it clear how the doctrine in the *Tesco* case relates to 'shadow directors'. Under company law, a shadow director is a person in accordance with whose directions or instructions the directors of a company are accustomed to act. Arguably, that description can apply to certain lenders; therefore one has to face the possibility that a lender who participates in the decision-making functions of a company might find itself liable for that company's crimes.

6.26 The liability of a company extends to the acts of its delegates. In the *Tesco* case, Lord Reid said:

> 'But the board of directors may delegate some part of their functions of management giving to their delegate full discretion to act independently of instructions from them.'

A company will not normally be liable for the actions of an independent contractor carrying out works for the company, for example a building contractor. However, liability may arise where the principal exercises complete control over the contractor, and joint liability may arise where control is shared or the principal exercises complete control over the contractor: see *RMC Roadstone Products Ltd v Jester* [1994] 4 All ER 1037.

Vicarious liability

6.27 In general, an employer is not directly liable for the crimes of his employee. Employer liability may arise, however, under statute, either expressly or by implication. Where such liability arises, it is described as 'vicarious liability'. If a statutory offence involves a mental element ('mens rea'), then vicarious liability may arise only where the employer has delegated the performance of a task to the employee: see *Vane v Yiannopoullos* [1974] 3 All ER 820 and *R v Winson* [1968] 1 All ER 197. Such delegation must be under the exclusive control of the delegate and free from the principal's supervision: see *Somerset v Hart* (1884) 12 QB 360. If the offence is one of strict liability, then vicarious liability may arise regardless of delegation, on the basis that the strict duty was the duty of the principal throughout: see *R v Winson* [1968] 1 All ER 197. In such a case, the employer remains liable notwithstanding that the employee may have disregarded his instructions, provided that the employee was acting in the course of his employment: see *Police Commissioners v Carman* [1896] QB 655. The employer is not liable where the employee does something entirely outside the course of his employment: see *Phelon and Moore Ltd v Keel* [1914] 3 KB 165.

An employer cannot be made vicariously liable for attempting or for aiding, abetting, counselling or procuring a crime: see *Gardner v Akeroyd* [1952] 2 All ER 306 and *Ferguson v Weaving* [1951] 1 All ER 412.

The *McAlpine* case

6.28 In *National Rivers Authority v Alfred McAlpine Homes East Ltd* [1994] 4 All ER 286, McAlpine's caused cement to pollute a stream contrary to s 85 of the WRA. Both the site manager and site agent accepted responsibility. McAlpine argued that a company could be made criminally liable only either by being vicariously liable for the acts of its servants or agents or by being identified with individuals representing the controlling mind and will of the company itself. It was accepted by the prosecution that neither the site manager nor the site agent could be considered as exercising that will. The Divisional Court held that McAlpine was liable, on the basis that the House of Lords decision in *Alphacell v Woodward* (see **6.9**) was an illustration of vicarious liability. This was said to be consistent with statements in subsequent cases that an employer was liable for pollution resulting from its operations carried out under its control, save only where some third party acted to interrupt the chain of causation. It did not matter that the company's head office had no direct part in the pollution incident. It was sufficient that those immediately responsible on site were employees of the company acting apparently within the course and scope of that employment.

The Divisional Court reviewed *Tesco Supermarkets v Nattrass* (see **6.24**) but did not follow it, because first, it did not involve strict liability (as did the *McAlpine* case) and secondly, because it involved the application of a statutory defence not the establishing of an offence. Morland J stated:

> 'In my judgment to make the offence an effective weapon in the defence of environmental protection a company must by necessary implication be criminally liable for the acts or omissions of its servants or agents during the activities being done for the company.'

Thus the *McAlpine* case means that a company may be liable for strict liability offences committed by its employees, but that the *Tesco* case can still be relied upon if a company is raising a statutory defence.

Executive liability

6.29 A director or senior manager who is involved in events leading up to or after a crime may, even though he did not participate in it, find himself liable as an accessory or for aiding, abetting, counselling or procuring the offence (see **6.20–6.21**). Furthermore, a number of statutes provide that if the offence was due to the negligence, consent or connivance of a director, manager, company secretary or other similar officer of the company, then that person can be liable for the offence. Typical statutes include the EPA (s 157), the WRA (s 217), the COPA (s 87) and the Health and Safety at Work Act 1974 (s 37). In certain circumstances, the Disqualification of

Directors Act 1986 provides that a person convicted of a criminal offence may be disqualified from acting as a director.

'Consent' means that the officer had knowledge of the relevant events: *Huckerby v Elliot* [1970] 1 All ER 189. There is no decided case on the meaning of 'connivance' in the context of this type of criminal provision. Presumably, its typical juxtaposition with 'consent' and 'neglect' means that it is not synonymous with either and, it is suggested, that it probably equates to wilful blindness. 'Neglect' implies failure to perform a duty of which the person knows or ought to know: *Re Hughes, Rea v Black* [1943] 2 All ER 269.

6.30 It is important to note that these statutory provisions also catch officers purporting to act as officers of the company – and this may include lenders or receivers who are in possession of corporate assets or acting as shadow directors. In *R v Boal* [1992] 3 All ER 177, it was held that the assistant general manager of a bookshop was not a 'manager' for the purposes of the Fire Precautions Act 1971, because he did not possess any real power or authority within the company.

PARTNERSHIPS AND UNINCORPORATED ASSOCIATIONS

6.31 A partnership, club or other unincorporated association is not treated as a separate legal entity. It is regarded simply as an association of separate legal persons, therefore any criminal acts committed by an unincorporated association are treated as the acts of responsible members. Thus, if a cricket club blocks a public footpath whilst playing its game, each and every responsible member could be prosecuted individually for obstruction of the footpath. Those actually blocking the path would be liable as principals and others could be liable as accessories or as aiding, abetting, counselling or procuring. If the criminal act is committed by an employee of the association, then individual members may find that they are vicariously liable for the actions of the employee.

PRIVATE PROSECUTIONS

6.32 Unless a statute restricts the right to prosecute to named bodies, then any member of the public may mount a private prosecution: see Prosecution of Offences Act 1985. As noted in Chapter 1, many pressure groups are now using this power.

CLASSIFICATION OF OFFENCES

6.33 There are three classes of offence, each characterised by mode of trial, namely (a) indictable offences, (b) summary offences, and (c) offences triable either way. The statute setting out the offence will prescribe which mode of trial applies. An offence triable on indictment can only be heard in the Crown Court. The hearing will take place in front of a judge and jury. The procedure commences in the magistrates' court, where the court must hold a preliminary inquiry to determine whether there is sufficient evidence to warrant a Crown

Court trial. An offence which is triable summarily will be heard in the magistrates' court only.

An offence triable either way can be heard either in the magistrates' court or the Crown Court. The prosecution is started in the magistrates' court with the laying of an information by the prosecution and the issue of a summons. The first hearing date will normally involve determining the mode of trial. The case can be tried by the magistrates' court if the defendant, prosecution and magistrates agree. Alternatively, the defendant can elect for Crown Court trial, or the magistrates may refer the matter to the Crown Court if they feel it is not suitable for them either because of its complexity or because their sentencing powers might be insufficient. The Magistrates' Court (Advance Information) Rules 1985 (SI 1985/601) provide that a defendant charged with an offence triable either way is entitled to information about the prosecution's case before the trial. This is an important right and the defendant should take advantage of it to obtain a full understanding of the case against him.

TIME-LIMITS

6.34 Section 127 of the Magistrates' Courts Act 1980 provides that an information alleging a summary offence must be laid within 6 months of the date of commission of the offence. There is no time-limit in relation to offences triable either way or on indictment only; however, undue delay in starting a prosecution can result in it being struck out for abuse of process.

PROVING THE FACTS

6.35 The prosecutor must, having overcome all the legal and procedural problems which face him, then be able to prove the facts which underpin the charge. This is, in the context of environmental pollution, a difficult task. Often everything can turn on the analytical chemist. If the offence involves the breach of objective standards, then the accuracy of his samples is paramount. If it involves judgemental tests, such as whether a substance is capable of causing harm, then he must be able to say, beyond all reasonable doubt, that such capability for harm existed in the circumstances of the alleged offence.

In 1994, Greenpeace was required to pay almost £29,000 in costs when prosecutions against ICI for alleged water pollution failed: see ENDS Report 234, p 47. The prosecutions failed because of difficulties over demonstrating that water samples were attributable to ICI and, if so, that they were accurate samples. If difficulties of this nature can arise in relation to evidence of direct breach of standards, then it is not difficult to see that the prosecution faces an uphill task in proving allegations that turn on secondary effects, such as disruption of the food chain or an ecosystem. It might appear that charges based on direct identifiable harm to flora and fauna or based on exceeding measurable standards are straightforward, but that is far from the truth. As noted in Chapter 13, laboratories can vary dramatically in their accuracy: see **13.28**.

Samples

6.36 Many statutes provide that samples are admissible in evidence only if a particular sampling procedure has been observed. Most common is the so-called 'tripartite' procedure, whereby the person taking the sample there and then separates it into three containers, which are then sealed, with one being delivered to the potential defendant: see s 209 of the WRA (see **7.19**). If the tripartite procedure is not required by statute, then the prudent prosecuting body will observe it none the less, in relation to samples, to reduce the possibility of a challenge at the trial.

6.37 Section 111(1) of the Environment Act 1995 will, when it comes into force, abrogate the sampling provisions in s 171(4) and (5) of the Water Industry Act 1991 and s 209(1), (2) and (4) of the WRA.

Chapter 7

WATER POLLUTION

INTRODUCTION

The nature of the problem

7.1 Water pollution is a source of concern for three reasons. First, such pollution may have an adverse effect on the flora and fauna which depend on the aquatic environment. Secondly, it may affect the natural beauty of that environment. Finally, it may adversely affect supplies of drinking water drawn from groundwaters. Assessing whether or not a particular site is polluting the aquatic environment, however, is not an easy task. In *Contaminated Land and the Water Environment* (Water Quality Series No 15; HMSO,1994) the National Rivers Authority states at para 4.2:

> 'Firstly, two of the most important indicators – the source and the receiving water – are often hidden from view and therefore quantifying or carrying out a risk assessment is a major operational undertaking. Similarly, because of other pollution sources – such as consented discharges, naturally-occurring sources of metals, or pollution incidents – attempts to quantify the percentage of pollution in a watercourse attributable to a contaminated land source is equally difficult. Seasonal and climatic fluctuations in the water cycle further complicate the situation.'

The extent of the problem is illustrated by Appendix 2(1), which lists the watercourses affected by discharges from abandoned coal mines and associated spoil tips. It also provides details of discharges from metal mines which have caused significant pollution problems.

This chapter will concentrate on the statutory sources of liability for water pollution, but it is important to recognise that common law liabilities may exist for water pollution, and interference with riparian rights, under negligence, nuisance, trespass and the rule in *Rylands v Fletcher*: see Chapter 5.

The National Rivers Authority

7.2 The Water Act 1989 was designed to restructure the water industry in line with the Government's drive towards privatisation, and the National Rivers Authority (NRA) was created by the Act to place the regulatory powers of the old water authorities into the hands of an independent body with no operational powers or duties. The NRA has a range of powers to deal with water pollution and s 2(2) of the Water Resources Act 1991 (WRA) places the NRA under a duty to promote the conservation and enhancement of the natural beauty and amenity of inland and coastal waters and associated lands, the conservation of flora and fauna dependant on an aquatic environment, and the use of such waters and lands for recreational purposes: see also ss 61A and 61B of the Land Drainage Act 1991 (as amended) for a similar duty in relation to internal drainage boards and local authorities. Section 3(2) of the WRA requires the NRA to protect against pollution any water which is, or is likely to be, abstracted. Section 19 requires the NRA to take all such actions as it may, from time to time, consider necessary or expedient, to conserve and secure the proper use of water resources.

7.3 The Government has announced that the NRA's powers and duties will be transferred to the Environment Agency in the future as a response, in part, to the problem of overlapping responsibilities between regulatory bodies. For example, the illegal deposit of waste which pollutes a watercourse may involve not only the NRA, but also the borough or district council (see Chapter 9) and the waste-regulation authority (see Chapter 8).

THE WATER RESOURCES ACT 1991

'Controlled waters'

7.4 The WRA is the principal statute for the regulation of water pollution. It is enforced by the NRA in respect of 'controlled waters', an expression which is defined at some length and complexity in s 104 and, in simple terms, covers the following:

(a) coastal and territorial waters (ie tidal waters and seawater within 3 nautical miles of the coastline);
(b) inland freshwaters; and
(c) groundwaters.

Land-locked waters that do not drain into controlled waters are excluded, as are certain waters prescribed by the Controlled Waters (Lakes and Ponds) Order 1989 (SI 1989/1149).

7.5 As to (b) above, 'inland freshwaters' are relevant lakes, ponds, inland rivers and 'watercourses': s 104 (1)(*c*). By s 221, a 'watercourse' includes

> 'all rivers, streams, ditches, drains, cuts, culverts, dykes, sluices, sewers and passages through which water flows except mains and other pipes which ... belong to the [WRA] or a water undertaker or ... are used by a water undertaker or any other person for the purpose only of providing a supply of water to any premises.'

This definition is subject to ss 72(2) and 113(1) in respect of certain sewers.

In *National Rivers Authority v Dovermoss Ltd* (1994) *The Times*, 8 February, the Court of Appeal held, by reference to the statutory extended meaning of 'watercourse', that 'controlled waters' can include dry watercourses or the overflow from flooded watercourses. The court decided that the words 'through which water flows' governed only the two words immediately preceding them, namely 'sewers' and 'passages'. They did not govern rivers, streams, ditches, drains and so on. Stuart-Smith LJ noted that ditches were often dry for a great part of the year but did not cease to be watercourses.

7.6 As to (c) above, groundwater is water which is contained in porous underground rocks and strata. If the groundwater flows through such rocks and strata, then they are known as 'aquifers' and can be significant sources of drinking water. About one third of the freshwater abstracted for public consumption in England and Wales comes from groundwaters.

The criminal offences

7.7 Section 85 of the WRA sets out six principal water pollution offences:

(a) causing or knowingly permitting any poisonous, noxious, polluting matter or solid waste matter to enter controlled waters;
(b) causing or knowingly permitting any matter, other than trade effluent or sewage effluent, to enter controlled waters from a drain or sewer in contravention of a prohibition under s 86;
(c) causing or knowingly permitting any trade effluent to be discharged into any controlled waters, or from land (through a pipe) outside the seaward limits of controlled waters;
(d) causing or knowingly permitting any trade effluent or sewage effluent to be discharged, in contravention of a prohibition under s 86, from a building or from any fixed plant onto or into any land or into any waters of a lake or pond which are not inland freshwaters;
(e) causing or knowingly permitting any matter whatever to enter any inland freshwater so as to tend (either directly or in combination with other matter) to impede the proper flow of the waters in a manner leading to, or likely to lead to, a substantial aggravation of pollution due to other causes, or the consequences of such pollution; and
(f) contravention of the conditions of a discharge consent.

The penalties for breach of s 85 are a maximum fine of £20,000 and/or 3 months in prison on summary conviction in the magistrates' court; or on conviction by a Crown Court, an unlimited fine and/or 2 years in prison. In *National Rivers Authority v Shell UK* (1990) *Water Law* 40, Shell was fined £1m for the escape of 30,000 gallons of crude oil into a watercourse. This was on top of a payment of £1.4m towards the costs of remedial operations.

7.8 In respect of (a), 'poisonous, noxious or polluting' means poisonous, noxious or polluting in relation to the watercourse as a watercourse and aquatic environment: see *Dell v Chesham District Council* [1921] 3 KB 427. In *National Rivers Authority v Egger (UK) Ltd* (1992) 209 ENDS Report 39, 4 LMELR 130, the defendant company was found guilty in respect of construction-site run-off, which had polluted the River Tyne at Hexham. The contaminant had a significant biochemical oxygen demand. There was no evidence of any fish kill or other damage, but the usefulness of the river as a watercourse had been reduced. In *National Rivers Authority v Dovermoss Ltd* (see **7.5**) the Court of Appeal held that proof of actual harm was not a prerequisite to liability for pollution under s 85: it was enough that there was a likelihood or capability of causing harm to animal or plant life or those who used the water. The *Egger* and *Dovermoss* cases concerned the word 'polluting' only; however, they indicate that evidence of actual harm is not necessary for liability in respect of 'poisonous' or 'noxious' matter. It should be enough for the prosecution to show that the substance was, by its nature, 'poisonous' or 'noxious'.

There is no definition of 'solid waste matter'; however, it is important to have regard to the word 'waste'. The tests discussed in Chapter 8 are relevant here; presumably, the deposit of solid materials which are not 'waste' would not fall within s 85. It is also worthy of note that the phrase 'solid waste matter' must encompass *inert* solid waste matter, thus liability may be incurred even though no actual damage to the watercourse or aquatic environment takes place – presumably on the basis that long-term solid discharges will eventually clog waterways and so on. It is essential to note that the NRA does not have to prove its case on the basis of impact on water quality or by reference to any prescribed environmental standards.

'Trade effluent' and 'sewage effluent' are defined in s 221(1) of the Act.

7.9 As to (b) and (d), s 86(1) provides that the NRA may serve notice on any person prohibiting him from making a discharge into controlled waters or requiring that any such discharge be subject to specified conditions. Section 86(2) provides that there is a breach of a prohibition if any discharge contains, or derives from a process involving, a 'prescribed substance'. The phrase 'prescribed substance' is not defined but s 221 defines 'prescribed' as meaning prescribed by regulations produced by the Secretary of State or the minister. On that basis, it would appear to refer to ss 82, 83 and 92, which deal with the Secretary of State's power to produce regulations in respect of water quality.

To 'cause' or 'knowingly permit'

7.10 It is an essential precondition to each of the s 85 offences that the defendant be shown to have 'caused or knowingly permitted' the relevant acts, and that phrase has been discussed in Chapter 6. Of particular relevance to s 85 are *Alphacell v Woodward*, *Wrothwell Ltd v Yorkshire Water Authority* and *National Rivers Authority v Welsh Development Agency* – they are discussed in **6.9**, **6.12** and **6.17** respectively. *National Rivers Authority v Alfred McAlpine Homes East Ltd* is important where pollution is caused by an employee: see **6.28**.

The offence of 'causing' under s 85 is one of strict liability, in the sense that it is not necessary for the prosecution to show intent or negligence on the part of the defendant: see *Alphacell v Woodward* and *Wrothwell Ltd v Yorkshire Water Company*. If the defendant was the effective cause of the pollution, then he will be liable. If, however, the pollution resulted from the act of another (eg a nocturnal trespasser) then the defendant should not be found guilty of this offence: see *Impress (Worcester) v Rees* [1971] 2 All ER 357. Likewise, he should not be found guilty of this offence if the offending substance is carried across or under his land, from another property, by natural causes, where his role is merely passive. The correct charge in the two last-mentioned situations might, if the necessary knowledge can be proved, be 'knowingly permitting': see **6.14–6.16**.

Executive and member liability

7.11 Section 217 of the WRA provides that if a corporation is guilty of an offence under the Act with the consent or connivance of or due to the neglect of any director, manager, secretary of similar officer or any person purporting to act in that capacity, then he will be guilty of the offence (see **6.23–6.30** generally). Section 217 also provides that, if the affairs of a corporation are managed by its members, then those members may also be criminally liable.

Statutory defences to section 85

7.12 Section 89 of the WRA provides some statutory defences to charges under s 85 including:

(a) the discharge was made in emergency to avoid danger to life or health, provided that the person responsible has taken all steps as are reasonably practicable in the circumstances to minimise the extent of the discharge and its polluting effects, and furnishes particulars of the discharge to the NRA as soon as is reasonably practicable thereafter;
(b) the discharge was of trade or sewage effluent from a vessel;
(c) the discharge was in waters from an abandoned mine; or
(d) if a discharge to inland freshwaters comes from solid refuse from a mine or quarry on any land and the deposit was made with the consent of the NRA, no other site for the deposit was reasonably practicable and the person responsible has taken all reasonably practicable steps to prevent the discharge.

As to defence (a), see *National Rivers Authority v ICI Chemicals and Polymers* (1992) 204 ENDS Report 37, 4 LMELR 131 and *Waste Incineration Services v Dudley Metropolitan Borough Council* (1992) Env LR 29. Section 89 also provides that a person shall not be guilty under s 85 by reason only of his permitting water from an abandoned mine to enter controlled waters. As we can see from Chapter 2 (at **2.16–2.17**), pollution arising from abandoned mines is a major problem. The defence in respect of abandoned mines in (c) will not apply to mines abandoned after 31 December 1999 when the relevant provisions of the Environment Act 1995 come into force: see **16.56**.

Section 88 provides defences where discharges are due to any act or omission pursuant to and in accordance with certain licences or consents, including discharge consents, granted by the NRA and waste-management or disposal licences. The statutory defence that the discharge was in accordance with an NRA consent is effective even if the permitted discharge involves a pollutant:

see *National Rivers Authority v Yorkshire Water Services* [1994] 2 WLR 1202. The burden of proving a statutory defence falls on the defendant.

Miscellaneous offences

7.13 There are, in addition to the WRA, numerous water pollution offences under other legislation. For example:

(a) section 68 of the Public Health Act 1875 in respect of pollution by gas producers;
(b) section 4 of the Salmon and Freshwater Fisheries Act 1975, concerning pollution which is poisonous or injurious to fish or the spawning grounds, spawn or food of fish;
(c) if the discharge is prejudicial to health or a nuisance, then it may be a statutory nuisance for s 79 of the Environmental Protection Act 1990;
(d) section 72 of the Water Industry Act 1991, in respect of the pollution of water in any waterworks to be used for human consumption; and
(e) section 38 of the Public Order Act 1986, in respect of the contamination of or interference with water.

National Rivers Authority's prosecution practice

7.14 It is important to note that the NRA will not necessarily prosecute for all breaches of s 85 – it may issue a caution instead. The NRA sets out three categories of offence:

- **Category 1** – where the pollution results in major or persistent environmental damage, which may include extensive fish kill, closure of abstraction, major remedial works and so on.
- **Category 2** – where the pollution has significant but not major results, which may include significant fish kill, notification to abstractors of water unfit for livestock, measurable effect on invertebrate life, contamination of river bed, or amenity value reduced by odour or appearance.
- **Category 3** – minor incidents of no notable effect or incapable of substantiation.

Category 3 incidents will normally result in a warning letter only. Category 2 incidents may involve a warning letter, a formal caution or prosecution depending on the facts of each case. Category 1 incidents will usually result in a prosecution. There is always room for negotiation where the incident is within Category 2. If the polluter has an otherwise good record on environmental matters, then it may be possible to end the matter with a caution or warning letter instead of a prosecution – especially if the defence can throw some doubt on the strength of the NRA's evidence. Any negotiations should, of course, be on a strictly off-the-record basis lest inadvertent admissions are later used in evidence. If it is agreed to settle on the basis of a caution, then the polluter will have to admit his guilt and the written note of the caution will form part of the polluter's record. As such, it is essential that the exact wording of any caution is agreed between the NRA and the polluter – in particular, the polluter should ensure that any mitigating factors are mentioned.

It is interesting to note that in 1991 the NRA reported 22,469 substantiated pollution incidents as against 356 prosecutions – the comparable figures for 1992 being 23,331 substantiated incidents and 297 prosecutions. The NRA issued 250 cautions in 1992 (*Water Pollution Incidents in England and Wales 1992*; NRA).

THE NATIONAL RIVERS AUTHORITY'S PREVENTATIVE AND RESTORATIVE POWERS

Section 161

7.15 Section 161 of the WRA provides that where it appears to the NRA that any poisonous, noxious, polluting matter or solid waste matter is likely to enter, or to be or have been present in, any controlled waters, then the NRA may carry out:

(a) where the matter appears likely to enter any controlled waters, works and operations to prevent it doing so; and/or
(b) where the matter appears to be or have been present in any controlled waters, works and operations –
  (i) to remove or dispose of the matter;
  (ii) to remedy or mitigate any pollution caused by its presence; or
  (iii) so far as is reasonably practicable, to restore the waters (including any flora and fauna dependent on the aquatic environment) to their state immediately before the matter became present in the waters.

The NRA has a discretion as to whether or not it will use its s 161 powers in any given case.

7.16 The NRA may then recover the expenses reasonably incurred in carrying out such works or operations from any person who caused or knowingly permitted:

(a) the matter to be present at the place where it was likely, *in the opinion of the NRA*, to enter any controlled waters; or
(b) the matter to be present in any controlled waters.

It is not clear whether or not the costs of surveys, tests and investigations relevant to the remedial works are recoverable under s 161(3).

Whilst the restoration powers in s 161 apply to abandoned mines, the power to recover expenses does not: s 161(4). Thus the NRA is incurring continuous restoration expenses in respect of the Wheal Jane tin mine in Cornwall whilst being unable to obtain reimbursement (see **2.16** above). Nor can the NRA impede or prevent any discharge under a discharge consent under the section: 161(2). Potentially, the s 161 liability for expenses could encompass not only owners or occupiers, but also lenders or receivers in possession.

There are no time-limits to the NRA's ability to recover its expenses and there is no reason why the ordinary 6-year time-limit for civil proceedings shall apply. Thus it is possible that the NRA could impose retrospective liability on erstwhile polluters – perhaps decades after the relevant incident.

The Environment Act 1995 provides amendments to s 161 which will, when they come into force, include a 'works notice' procedure: see **16.57**.

Section 161 – the drawbacks

7.17 The main drawback with s 161, from the NRA's standpoint, is that it has to incur remediation costs before trying to recover them from the alleged polluter. There are three immediate problems. First, the NRA has limited and apparently shrinking resources. In 1991/92 its 'grant-in-aid' funding from government was £106m. It was then predicted that the figure for 1994/95 would be £82m; however, that figure was reduced to £68m in 1993. It is anticipated that it will fall to £60m for 1995/96 and £58m for 1996/97. In 1990 the NRA asked the Department of the Environment for £34m over 3 years to fund remedial works, but the bid was rejected with the suggestion that the NRA explore other funding possibilities: see *Contaminated Land and the Water Environment* (Water Quality Series No 15; HMSO, 1994). In November 1993, Lord Crickhowell, the NRA's chairman, said that the economies being demanded by the Government 'can only be made by significantly reducing the level of service the NRA provides, or by substantially increasing the charges we make'. The 1994/95 corporate plan noted that the NRA had shed 736 posts during 1993/94 and that further reductions would take place in the future.

Secondly, the NRA cannot guarantee that it will recover its expenses. The persons made liable by s 161 may have vanished or become bankrupt. Even if those persons can be identified, they are unlikely to pay up without a fight – both in relation to liability and as to the amount of the expenses. Protracted litigation can involve massive costs and such costs would be awarded in favour of the NRA only if it is successful in the litigation. If it loses, then the NRA would normally be liable for the defendant's costs.

Finally, bringing water resources back to acceptable levels may take decades, and the NRA does not see itself in the business of long-term site management. The NRA sees s 161 as providing 'no more than a fire-fighting power': see Water Quality Series No 15 above, para 5.7.

Section 161 in practice

7.18 Clearly, there are considerable problems in the way of implementing s 161; however, it would be wrong to pretend that it has little practical importance – indeed the reverse is true. The NRA is well aware of the fact that most businesses are very sensitive to their public images and, as such, will often come to a settlement provided that publicity is avoided. Furthermore, incidents which fall within s 161 often fall within s 85 as well; therefore the NRA will have the leverage of a pending criminal prosecution behind its requests for the expenses of restoration.

SAMPLING REQUIREMENTS

7.19 Section 209 of the WRA provides that if, in any legal proceedings, the

NRA wishes to rely on the result of the analysis of any sample taken by it, then its officer must have:

(a) notified the owner or occupier of its intention to have the same analysed;
(b) there and then divided the sample into three sealed and marked containers;
(c) delivered one container to the owner or occupier; and
(d) retained one container for future comparison.

If it is not reasonably practicable for the officer to comply with the above steps, then he must take them as soon as reasonably practicable after the sample was taken: s 209(2).

In *Attorney-General's Reference (No 2 of 1994)* [1994] 1 WLR 1579, the Court of Appeal ruled that the division of the sample should take place at or proximate to the site and on the occasion of taking the sample. The division did not need to take place at the exact moment and exact spot that the sample was taken. The court also ruled that the NRA did not have to give the owner or occupier advance notice that a sample was about to be taken. It is interesting to note that the officer's duty to provide a copy sample extends to the owner or occupier only. It will be recalled that ss 85 and 161 extend liability to any person causing or knowingly permitting water pollution (eg employees, contractors, lenders and receivers in possession), yet these potential defendants are not (probably for practical reasons) included. It follows that any contractual or leasehold arrangements should ensure that any samples given to owners or occupiers are held in trust for other interested parties.

The procedure in s 209 does not apply to the monitoring of water quality by probes or static testing devices because no 'sample' is actually taken by such mechanisms: see *CPC (UK) v National Rivers Authority* (1994) 235 ENDS Report 41.

Section 111 of the Environment Act 1995 will, when it comes into force, abrogate the sampling requirements in s 209(1), (2) and (4).

THE NATIONAL RIVERS AUTHORITY'S POWERS OF ENTRY

7.20 Section 169 of and Sch 20 to the WRA give the NRA's officers powers of entry to private land to ascertain whether or not any of the water pollution provisions of the Act have been contravened. The power includes the right to carry out experimental borings or other works on the premises and to install and keep monitoring and other apparatus there.

In the majority of cases, the power of entry may be exercised at any reasonable time unless there is an emergency. The NRA officer need not give advance notice unless the premises are in residential use or the entry is to be with heavy equipment.

7.21 The officer must have a written designation of his right to use the power of entry and must produce that designation, and any other authority, before he exercises the power. The NRA's officer may obtain a warrant to exercise the power from a justice of the peace if:

(a) entry has been refused;
(b) refusal is reasonably anticipated;
(c) the premises are unoccupied or the occupier is temporarily absent;
(d) the case is one of urgency; or
(e) application for entry would defeat the object of the proposed entry.

Any person who intentionally obstructs another person acting in the exercise of a power of entry is guilty of a criminal offence.

THE GROUNDWATER DIRECTIVE

7.22 In December 1979 the EC issued Directive 80/68/EEC in respect of the pollution of groundwater by toxic substances – the date for compliance being December 1981. The Directive provides that the discharge of 'List I' substances must be prohibited and 'List II' discharges must be controlled. These provisions also relate to activities which may cause indirect discharges. In July 1992, the NRA assumed responsibility for the administration of the National Classification Scheme for dangerous substances under the Directive. A National Advisory Group was set up to advise on substances which should be placed on Lists I and II respectively.

The EC itself is not happy with the Directive: it has been criticised for lack of technical detail and the EC Fifth Action Programme on the Environment observed that it was not achieving its objectives. In 1991, EC ministers attended a seminar at The Hague which made it clear that modern industrial and farming practices posed a considerable threat to the protection of groundwaters. In February 1992, the EC Council passed a resolution calling on the Commission to revise the Directive and prepare an action programme. On 4 October 1994, the environment ministers of the EC adopted a resolution urging the Commission to speed up its work on groundwater noting that groundwaters were threatened in a number of areas in the EC.

Chapter 8

WASTE MANAGEMENT

INTRODUCTION

8.1 It is necessary to discuss waste regulation for two main reasons: first, because land is often contaminated by waste (or contaminants deriving from waste) on or in that land or adjacent land; secondly, because remedial works will often give rise to waste or involve the treatment or transportation of waste. There are comprehensive statutory controls on the deposit, transportation and handling of controlled waste. These controls may apply to the original owner, occupier or operator and may also apply to lenders, lessees and persons involved in later transactions in the land.

THE REGULATORY FRAMEWORK

8.2 The principal regulatory regime for the recovery and disposal of waste is, from 1 May 1994, the system created by Part II of the Environmental Protection Act 1990 (EPA) and administered by the waste regulation authorities. In England, the authorities will normally be either the county council or the metropolitan district council and, in Wales, the district council. There are separate authorities for Greater London, Greater Manchester and Merseyside. As from April 1996 (on the current timetable), the Environment Agency will assume the responsibilities of the waste regulation authorities. The new system is designed to transpose the requirements of the EC on waste management into UK law and practice. This has not proved to be an easy task and has resulted in an extremely complex arrangement of primary and secondary legislation which starts by redefining the EPA's definition of 'controlled waste'.

'Controlled waste'

8.3 The EPA regulatory regime is designed to deal with 'controlled waste', namely 'household, industrial and commercial waste or any such waste': s 75(4). 'Household waste' is, broadly, waste from a dwelling or residential home or premises forming part of a hospital, nursing home, university, school or other educational establishment or a residential caravan on a caravan site: s 75(5). 'Industrial waste' is waste from any factory within the meaning of the Factories Act 1961, including land occupied by certain public or transportation bodies: s 75(6). 'Commercial waste' is waste from premises used wholly or mainly for trade, business, sport, recreation or entertainment but excluding household and industrial waste, and mine, quarry and agricultural waste: s 75(7).

'Directive waste'

8.4 It is now necessary to pick up a fundamental point arising from the foregoing discussion, namely the meaning of the expression 'waste'. Section 75(2) of the EPA provides the primary definition of 'waste' as including:

> '(a) any substance which constitutes a scrap material or an effluent or other unwanted surplus substance arising from the application of any process; and,
> (b) any substance or article which requires to be disposed of as being broken, worn out, contaminated or otherwise spoiled.'

'Waste' does not include explosives or sewage. Anything which is discarded or otherwise dealt with as if it were waste shall be presumed to be waste unless the contrary is proved: s 75(3).

The definition in s 75(2) is due to be amended under the Environment Act 1995 (see **16.60**) and the presumption in s 75(3) will then cease to have effect (see **16.61**).

8.5 That definition does not coincide with the definition of waste (which is known as 'Directive waste') stipulated by EC Directive 75/442 (referred to in this chapter as 'the Directive'), as amended by Directives 91/156 and 91/692. The Government has, in fact, replaced the definition of waste in s 75(2) of the EPA with the EC's definition, but without repealing s 75(2)! This feat has been achieved by way of the Waste Management Licensing Regulations 1994 (SI 1994/1056). Effectively, the regulations inch the s 75(2) definition of 'waste' around to refer, in practice, to Directive waste exclusively. There are two fundamental modifications. First, s 75(2) has to be read as including Directive waste within its definition of waste. Secondly, anything which is not Directive waste is treated as not being controlled waste. The regulations include a sweeping-up provision to the effect that any reference to 'waste' in Part II of the EPA includes a reference to Directive waste. Effectively, therefore, 'controlled waste' and 'Directive waste' are now one and the same.

Is it Directive waste?

8.6 The next question to tackle is the practical one, namely when is a particular substance or material Directive waste? Unfortunately, not only is the answer not a simple one but it is also not an answer which can be provided in full within the scope of this book. The many complexities of the topic are the subject of Circular 11/94 – a massive (and probably unprecedented) 199 pages of advice from the Department of the Environment (DoE). The following discussion must, perforce, be restricted to an outline sketch only.

Paragraph 2.18 of Circular 11/94 advises that in order for something to be Directive waste it must:

(a) fall within Part II to Sch 4 of the 1994 regulations; and
(b) fall within one of the following categories:

(i) be discarded, disposed of or got rid of by the holder;
(ii) be intended to be discarded, disposed of or got rid of by the holder; or
(iii) be required to be discarded, disposed of or got rid of by the holder.

Test (a) and one of the three tests in (b) must be fulfilled. As to (a), Part II of Sch 4 to the regulations is set out in Appendix 2(2) and it will be seen that it is designed to correspond with the EC Directive. As to the meaning of 'discard', see **8.17**.

8.7 The tests in (b) are extremely problematic for a number of reasons – not least because what is discarded by one man may become another man's raw material as opposed to being mere 'rubbish'. The Circular states (para 2.14):

> 'Waste appears to be perceived in the Directive as posing a threat to human health or the environment which is different from the threat posed by substances or objects which are not waste ... It leads the Departments to the view that the purpose of the Directive is to treat as waste ... those substances or objects which fall out of the commercial cycle or out of the chain of utility'

and then adds (para 2.15):

> '... the purpose of the Directive is to supervise the collection, transport, storage, recovery and disposal of waste; and these are activities which themselves may be of a commercial nature. A distinction must be drawn, therefore, between the normal commercial cycle and the commercial cycle which exists for the purpose of collecting, transporting, storing, recovering and disposing of waste.'

8.8 The fact that a material is surplus to the producer's requirements does not make it waste if that material is taken and used by another person without altering its nature or form (para 2.35 of the Circular), but the recycling of waste falls squarely within the waste management regime. In order to explain the position of recycling, the Circular coins the phrase 'specialised recovery operation', meaning an operation which wholly or partly derives its justification from the recovery of waste: see para 2.30. If a degenerated substance or object can be put to use only by an establishment or undertaking specialising in waste recovery, even if for value, then the Circular treats it as being no longer part of the normal commercial cycle or chain of utility and, therefore, waste: para 2.33(c)(i).

8.9 In January 1994, the European Commission issued the European Waste Catalogue, which lists materials qualifying as waste with the objective of harmonising definitions of waste across Member States. The list is not exhaustive, and inclusion of a material within the Catalogue does not necessarily mean that it is waste for all purposes – it is waste only when it meets the tests in the Directive: see ENDS Report 228, p 42.

Excluded classes of waste

8.10 Regulation 1(3) of the 1994 regulations states that the definition of Directive waste does not include anything excluded by Art 2 of the Directive, namely:

(a) certain gaseous effluents emitted to the atmosphere from non-waste processes;
(b) where they are already covered by other legislation –
  (i) radioactive waste;
  (ii) certain waste resulting from mineral and quarry workings;
  (iii) animal carcasses and certain agricultural waste;
  (iv) waste waters (with the exception of waste in liquid form); and
  (v) decommissioned explosives.

As to the words 'where they are already covered by other legislation', the Secretary of State is of the view that the five classes of waste are already covered by existing UK and EC legislation: see p 13 of Circular 11/94. These exclusions are similar to those already in the EPA and the Secretary of State has indicated that the differences between the two regimes will be adjusted by further regulations.

Exempt classes of waste

8.11 Schedule 3 to the 1994 regulations lists 43 activities which are exempt from waste management licensing but which, if carried out by any establishment or undertaking after 31 December 1994, must be registered with the appropriate authority: see Art 18. Paragraphs 9 and 19 of the Schedule exempt certain types of storage and re-use of waste in regeneration and construction works. Subject to certain conditions, para 15 exempts, somewhat ambiguously, 'the beneficial use of waste' if it is put to that use without further treatment and the use does not comprise an operation within Part III of Sch 4. The Schedule also exempts a wide range of recycling activities. Article 16 of the regulations provides a list of activities which are excluded both from waste management licensing and registration under Art 18 because they are subject to other control regimes.

Special waste

8.12 Section 17 of the Control of Pollution Act 1974 (COPA) (now s 62 of the EPA) gives the primary definition of 'special waste' as controlled waste (now 'Directive waste') of any kind which is or may be so dangerous or difficult to dispose of, that special provision is required for disposing of it. Further definition is given by the Control of Pollution (Special Waste) Regulations 1980 (SI 1980/1709) and advice on the classification is given by Waste Management Paper 23, *Special Wastes – A Technical Memorandum Providing Guidance on their Definition*' (HMSO, 1981). These provisions implement EC Directive 78/319 on toxic and dangerous waste. The 1980 regulations require that special-waste consignments be notified to the waste disposal authorities, that prescribed consignment notes be used on regular consignments and that the location of deposits of special waste be recorded. The Government is currently reviewing the 1980 regulations in light of certain EC developments and the DoE issued a consultation paper on the topic, *Special Waste and the Control of its Disposal*, in February 1990.

8.13 Section 63(2) of the EPA purports to create criminal liability for the deposit of waste other than controlled waste in any case where:

> 'if the waste were special waste and any waste management licence were not in force, he would be guilty of an offence under section 33.'

Unfortunately, the problem with s 63 is that special waste is, by definition, controlled waste. Presumably, the section is directed towards non-controlled waste which displays the characteristics of special waste – but the position is far from clear. Section 63 will be amended when para 81 of Sch 22 to the Environment Act 1995 comes into force.

THE WASTE MANAGEMENT REGIME

8.14 Section 33(1)(a) of the EPA provides that, subject to certain exceptions, a person shall not deposit controlled waste (ie Directive waste) or knowingly cause or knowingly permit it to be deposited in or on any land unless a waste management licence authorising the deposit is in force and the deposit is in accordance with the licence. Section 33(1)(b) requires that a licence must also be in force before the treating, keeping or disposing of controlled waste (a) in or on any land (a 'site licence') or (b) by means of any mobile plant (a 'mobile plant licence') or knowingly causing or knowingly permitting these activities. At present, only mobile incinerators may be licensed under s 33(1)(b): see Art 12 of the 1994 regulations. Current waste disposal licences granted under COPA are treated as equivalent to waste management licences under the EPA: see s 77(2).

8.15 Section 33(1)(c) provides that it is an offence to treat, keep or dispose of controlled waste in a manner likely to cause pollution of the environment or harm to human health. A prosecution under s 33(1)(c) is not precluded because there is a waste management licence in force for the site.

8.16 In order to bring s 33 into line with the Directive, para 9 of Sch 4 to the 1994 regulations states that s 33(1)(a) and (c) are to be read as including the operations listed in Parts III and IV of Sch 4 to the regulations (see Appendix 2(2)) but that the effect of s 33(1)(b) is limited to those operations only.

8.17 Under s 29(6), waste is 'treated' when:

> 'it is subjected to any process including making it re-useable or reclaiming substances from it and "recycle" (and cognate expressions) shall be construed accordingly.'

Taking the foregoing provisions as a whole, it is clear that reference must be made to the lists in Sch 4 before engaging in any remedial treatment to a site, because that treatment may require a waste management licence. For example, para 5 of Part III refers to 'specially engineered landfill of waste' and includes the placement of waste into lined and capped cells. There is nothing in the provision limiting it to waste brought freshly onto the site, therefore it is arguable that it applies to redistribution of waste already on the site. In that context, the author has had sight of a joint opinion from Mr A. Scrivener QC and Mr J. Harper QC (by kind courtesy of Camerons solicitors of London WC2) as to whether or not a licence was needed for the extraction of contaminated material from the ground within a site owned by British Gas in Sydenham, which was then redistributed over the site as part of a ground-levelling exercise. The advice turned on the meaning of the word 'discard' in the definition of 'Directive waste' in the 1994 regulations. Both counsel came to the view that 'discard' includes 'deposit' and that a licence was needed. Paragraphs 2, 8 and 9 of Part III and paras 1, 2, 3 and 4 of Part IV need to be considered before treating land contaminated by waste with biological or physico-chemical treatments.

8.18 Any person making arrangements for the disposal or recovery of controlled waste on behalf of the holder needs to be aware of Art 18 of the 1994 regulations, which requires that dealers and brokers in controlled waste must be registered.

8.19 Section 33 creates liability for many people who are not, and never will be, waste disposal or landfill specialists. It is essential that the waste regulation authorities are notified and their advice followed in relation to potential licensing matters. This is an extremely complex topic and the unwary owner or developer may, in all innocence, find himself in trouble. For example, X might decide to remove some controlled waste from his site A to his nearby site B, perhaps only a matter of yards, in order to develop site A. He will need a waste management licence because he will be depositing or disposing of the waste. He may still need a licence if, instead, he decides merely to move the waste from one part of site A to another part of the same site.

8.20 Failure to obtain a waste management licence when it is required or to observe the conditions of such a licence is a criminal offence triable either way and punishable with a maximum fine on summary conviction of £20,000 or up to 6 months' imprisonment or both, and on Crown Court conviction by an unlimited fine or 2 years' imprisonment or both: s 33(8).

Temporary deposits

8.21 In *Leigh Land Reclamation Limited v Walsall Metropolitan Borough Council* [1991] JPL 867 the High Court took the view, in the context of s 3 of COPA, that the word 'deposit' indicated that the waste in question must have been finally laid to rest – thereby excluding waste which had been placed on land temporarily with a view to a later transfer elsewhere. However, in *R v Metropolitan Stipendiary Magistrates ex parte London Waste Regulation*

Authority and Others [1993] 3 All ER 113, the High Court took the view that the notion that the term 'deposit' necessarily involves a final resting place is mistaken and expressly overrode the *Leigh Reclamation* case.

Both decisions were by the Divisional Court, therefore they are of equal status; however, it is submitted that the *Leigh Reclamation* case was wrongly decided. Section 4(3)(a) of COPA allowed the Secretary of State to exclude from control, by regulations, deposits 'of such a *temporary* nature that they may be so excluded' (emphasis added). That provision is repeated in s 33(4)(a) of the EPA. The mere fact that ss 4(3)(a) and 33(4)(a) exist at all demonstrates, it is submitted, that the draftsman sought to cover both temporary and permanent deposits under the principal provisions. If the situation were otherwise, then ss 4(3)(a) and 33(4)(a) would be otiose.

8.22 It should be noted that certain types of temporary storage of waste pending transportation are exempted from licensing by paras 40 and 41 of Sch 3 to the 1994 regulations and excluded by para 15 of Part III and para 13 of Part IV to Sch 4.

Defences to section 33 offences

8.23 It is a defence for a person charged with an offence under s 33 to prove:

(a) that he took all reasonable precautions and exercised all due diligence to avoid the commission of the offence;
(b) that he acted under instructions from his employer and neither knew nor had reason to suppose that the acts done by him constituted a contravention of s 33(1); or
(c) that he acted in an emergency in order to avoid danger to the public and that, as soon as reasonably practicable after the relevant actions were done, particulars of them were furnished to the waste-regulation authorities.

As to (a), see *Austin Rover Group v Her Majesty's Inspector of Factories* [1989] 3 WLR 520 at p 527 and *Riverstone Meat Co Pty v Lancashire Shipping Co* [1960] 1 All ER 193 at p 219. The defence in (c) will be amended when para 64 of Sch 22 to the Environment Act 1995 comes into force.

Civil liability for unlawful deposits

8.24 Section 73 of the EPA provides that if controlled waste is deposited in or on land so as to cause an offence under s 33(1) or s 63(2) of the EPA to be committed, then, subject to stipulated exceptions, any person who knowingly caused or knowingly permitted the deposit is, prima facie, liable for any damage caused by that deposit. This provision is discussed more fully in Chapter 5 (see **5.34**).

THE OBSERVANCE OF LICENCE CONDITIONS

8.25 Section 42 of the EPA places a general duty on the waste regulation authorities to ensure that the conditions of licences are complied with and that activities authorised by licences do not cause pollution of the environment, harm to human health or serious detriment to the amenities of the affected locality. If a licence holder fails to comply with the conditions of the licence, then the authorities may serve a notice on him requiring compliance and, in default, may revoke or suspend the licence: s 42(5) and (6). The authority may, in an emergency, carry out work on the site and recover the expenditure from the present or a former licence holder: s 42(3) and (4).

Section 35(4) of the EPA

8.26 Section 35(4) contains what appears, at first reading, to be a remarkable provision:

> 'Conditions may require the holder of a licence to carry out works or do other things notwithstanding that he is not entitled to carry out the works or do the thing and any person whose consent would be required shall grant, or join in granting, the holder of the licence such rights in relation to the land as will enable the holder of the licence to comply with any requirements imposed on him by the licence.'

This provision creates considerable uncertainty for adjoining owners or occupiers. Taken literally, the provision would entitle licence holder A to enter adjacent landowner B's site in order to discharge a licence condition whether or not B had ever been involved in the grant of the licence or objects to A's entry onto his land. There is nothing to indicate that B would be compensated for this imposition regardless of the damage done to his land. Nor is there anything which requires that the adjacent or otherwise affected landowner be notified that such a condition is proposed or allowing him to appeal against such a condition.

NOTICES TO REMOVE WASTE

Section 59 of the EPA

8.27 If controlled waste is deposited in breach of the licensing provisions of the EPA, then the waste regulation authority or a collection authority may serve a notice on the occupier of the land requiring him to remove the waste or take steps with a view to eliminating or reducing the consequences of the deposit: EPA, s 59.

8.28 Section 59(2) provides rights of appeal to the magistrates' court and the court may quash the notice if the occupier neither deposited the waste nor knowingly caused or knowingly permitted the deposit or there is a material defect in the notice. An appeal must be filed within 21 days of service of the s 59 notice.

The word 'occupier' is not defined by the EPA and it could be argued that the term includes leaseholders, mortgagees in possession and, perhaps, receivers. Nor does the section indicate that the occupier served with the s 59 notice must have been the occupier at the time of the deposit; in other words the notice is served on the current occupier.

8.29 Section 59(5) and (6) provide that if the recipient fails, without reasonable excuse, to comply with the notice then:

(a) he may be fined up to level 5 on the standard scale with a further daily fine of up to one-tenth of level 5 for each day that the default continues; and
(b) the authority may do what that person was required by the notice to do and may recover from him any expenses reasonably incurred in doing so.

8.30 Section 59(7) then goes on to give waste regulation authorities and collection authorities a further remedial power which may be exercised summarily and without giving prior notice to the occupier (if any). The power may be exercised if it appears to the relevant authority that waste has been deposited in breach of s 33(1) and:

(a) in order to remove or prevent pollution of land, water or air or harm to human health it is necessary that the waste be forthwith removed or other steps taken to eliminate or reduce the consequences of the deposit or both; or
(b) there is no occupier of the land; or
(c) the occupier neither made nor knowingly permitted the deposit of the waste.

The authority may remove the waste from the land or take other steps to eliminate or reduce the consequences of the deposit, and may, under s 59(8), then recover the necessary costs of doing so and of disposing of any waste removed in the exercise of the power under s 59(7):

(a) in a case falling within (a) above, from the occupier unless he proves that he neither made, knowingly caused nor knowingly permitted the deposit; or
(b) in any other case, from any person who deposited, knowingly caused or knowingly permitted the deposit of any of the waste.

The power can, thus, catch any person in the position of being able to 'knowingly cause' or 'knowingly permit' (discussed in Chapter 6) the deposit, and is not restricted to occupiers only. Thus, a landlord may be caught for these costs – as could a lender who is exercising control over the site. Furthermore, the reference to 'any of the waste' in (b) appears to create joint and several liability where there is more than one depositor.

Section 59: the drawbacks

8.31 One of the most significant problems with s 59 is that it does not apply to deposits which predate the regulation of controlled waste by COPA, in other words before 14 June 1976. Thus the first task for the waste-regulation authority is to obtain evidence of the date of deposit, and that can be an almost impossible task in many cases – especially with mixed deposits.

8.32 Another problem is that the s 59 notice has to be served on the current occupier of the relevant land, yet he will be able to escape liability on appeal to the magistrates' court if he was not involved in the original deposit – for example, he may not have been the occupier at that time. Again, the authority is faced with the problem of establishing the date and circumstances of the deposit, failing which it has to rely on its default power under s 59(7).

THE 'DUTY OF CARE'

8.33 Developing a site will often involve removing waste from it and it is important to have regard to the 'duty of care' under s 34 of the EPA before doing so. Section 34 was brought into effect in part on 1 April 1992. It imposes a duty of care on any person who imports, produces, carries, keeps, treats or disposes of controlled waste or, as a broker, has control of such waste to take all such measures as are reasonable in the circumstances:

(a) to prevent any contravention by any other person of s 33 of the EPA;
(b) to prevent the escape of the waste from his control or that of any other person; and,
(c) on transfer of the waste, to secure that the transfer is to an authorised person or to a person for authorised transport purposes and that there is transferred such a written description of the waste as will enable others to avoid contravention of ss 33 and 34(1)(b).

An 'authorised person' includes a waste collection authority, the holder of a waste management or waste disposal licence, a registered carrier of controlled waste or any person exempted by regulations under the Control of Pollution (Amendment) Act 1989: s 34(3). A new s 34(3A) will be added to the EPA when para 65 of Sch 22 to the Environment Act 1995 comes into force, which will empower the Secretary of State to amend s 34(3) by regulations.

With effect from 3 November 1994, s 33 of the Deregulation and Contracting Out Act 1994 inserts a new s 33(4A) in the EPA which provides that:

(a) a transfer of waste in stages shall be treated as taking place when the first stage of the transfer takes place; and
(b) a series of transfers between the same parties of waste of the same description shall be treated as a single transfer taking place when the first of the transfers in the series takes place.

A producer will not be liable for the acts or omissions of an employee where he has taken all reasonable steps to ensure that the duty of care is observed: see *Seaboard Offshore Ltd v Secretary of State for Transport* [1994] 2 All ER 99.

8.34 Additionally, the Environmental Protection (Duty of Care) Regulations 1991 (SI 1991/2839) require that a 'transfer note' be completed at the same time as the written description required by s 34(1) is transferred and then kept for two years by the transferor and transferee. The Secretary of State for the Environment has produced a Code of Practice and, by s 34(7) and (10), it is admissible in court proceedings: see also DoE Circular 18/91.

It is not clear whether or not s 34(1)(c) and the 1991 regulations create two independent offences which can be charged concurrently or whether, alternatively, the 1991 regulations are parasitic in that they depend on the transfer of the written description for s 34(1)(c) before they apply. The regulations clearly state that the transfer note must be completed 'at the same time as the written description of the waste is transferred' and this appears to be the key to the operation of the requirement to complete and sign a transfer note. The pertinent question, therefore, is 'what happens where no written description is transferred?' Arguably, it is the transfer of the written description which triggers the requirement for the transfer note and a complete failure to transfer a written description would appear to frustrate the operation of the 1991 regulations. It was open to the Secretary of State, in drafting the regulations, to provide that transfer notes should be completed on the transfer *of controlled waste* itself: but he did not do so and this suggests that he deliberately tied the 1991 regulations to the transfer of the written description only. If this is right, then a waste regulation authority cannot prosecute for breaches of both s 34(1)(c) and the 1991 regulations in relation to the same transfer of controlled waste. If no written description was transferred, then, on this approach, only a breach of s 34(1)(c) can be charged.

8.35 Failure to comply with the duty of care or the 1991 regulations is an offence triable either way and punishable on summary conviction by a maximum fine not exceeding level 5 on the Standard Scale or by an unlimited fine in the Crown Court: s 34(6).

Application and duties

8.36 The duty of care is surprising in its range of application and the duties it imposes. For example, if a developer demolishes and removes a building from a site, then he must ensure that the contractor removing the waste is authorised to do so and that the appropriate written documentation is made out. As to the waste carried, para 3.4 of the Code of Practice states:

> 'Anyone intending to transfer waste to a carrier will need to check that the carrier is registered or exempt from registration. A registered carrier's authority for transporting waste is either his certificate of registration or a copy of his certificate of registration if it was provided by the waste regulation authority ... Photocopies are not valid and do not provide evidence of the carrier's registration.'

Paragraph 3.5 adds:

> 'In all cases other than those involving repeated transfers of waste, the holder should ask to see, and should check the details of, the carrier's certificate or copy certificate of registration ... In addition, before using any carrier for the first time, the holder should check with the waste regulation authority with

which the carrier is registered that his registration is still valid, even if his certificate appears to be current.'

8.37 The developer (ie the 'producer' of the waste for the purposes of the Code of Practice) must also do what is reasonable in the circumstances to ensure that the disposal site is licensed and that the waste does not escape on its trip there. Paragraph B7 of the Code of Practice states:

> 'It is not possible to draw a line at the gate of producers' premises and say that their responsibility for waste ends there. A producer is responsible according to what he knows or should have foreseen. So if he hands waste to a carrier not only should it be properly packed when transferred, but the producer should take account of anything he sees or learns about the way in which the carrier is subsequently handling it. The producer would not be expected to follow the carrier but he should be able to see whether the waste is loaded securely for transport when it leaves, and he may come to learn or suspect that it is not ending up at a legitimate destination. A producer may notice a carrier's lorries returning empty for further loads in a shorter time than they could possibly have taken to reach and return from the nearest lawful disposal site; or a producer may notice his carrier apparently engaged in the unlawful dumping of someone else's waste. These would be grounds for supposing illegal disposal of his own waste. The same reasoning applies when a producer makes arrangements with a waste manager for the treatment or disposal of waste. The producer shares the blame for illegal treatment of his waste if he ignores evidence of mistreatment; a producer should act on knowledge to stop the illegal handling of waste.'

8.38 On 18 August 1995 the Department of the Environment published a draft revised Code of Practice for consultation. Copies can be obtained from the Department at: Romney House, 43 Marsham Street, London, SW1P 3PY.

Chapter 9

STATUTORY NUISANCE

INTRODUCTION

9.1 'Statutory nuisance' has existed, in one form or another, for well over 100 years. Its current form is in Part III of the Environmental Protection Act 1990 (EPA), which came into force for these purposes on 1 January 1991. Statutory nuisance gives local authorities and certain members of the public considerable power to abate nuisance in a variety of environmental media, including smells, smoke, dust, stream, effluvia and noise. 'Abatement notices' may be served or 'abatement orders' obtained which not only require the abatement of the nuisance but also entitle the relevant local authority to abate the pollution and recover its expenses from a variety of persons, who may include the polluter or the owner or occupier of the offensive premises.

9.2 The remedies for statutory nuisance are very wide-ranging and flexible. They are normally enforced through the magistrates' courts and therefore are relatively speedy. The first formal step is the service of an abatement notice by the local authority, or an abatement order by the magistrates' court, requiring abatement of the nuisance. If the recipient fails to comply with the notice, then it may be enforced through the magistrates' court by way of fines, including daily penalties. In the extreme case, the local authority may obtain an injunction in the High Court.

The statutory nuisance provisions create liability not only for freeholders and occupiers of land, but also for lenders and receivers in possession. As such, they are of wide-ranging concern.

'PREJUDICIAL TO HEALTH OR A NUISANCE'

9.3 Section 79 requires local authorities to inspect their areas for 'statutory nuisances'. Such nuisances arise where premises are 'prejudicial to health or a nuisance' by virtue of (inter alia):

(a) the state of the premises (s 79(1)(a));
(b) certain types of smoke emitted from premises (s 79(1)(b));
(c) fumes or gases emitted from private dwellings (s 79(1)(c));
(d) any dust, steam, smell or other effluvia arising on industrial, trade or business premises (s 79(1)(d));
(e) any accumulation or deposit (s 79(1)(e));
(f) noise emitted from premises (s 79(1)(g)); or
(g) any other matter declared by an enactment to be a statutory nuisance (s 79(1)(h)).

Items (a), (d), (e) and (g) are of particular concern in relation to contaminated land. As to (d), see the note at **9.29** below on 'industrial, trade or business premises'.

As to (g), it is noted that s 259 of the Public Health Act 1936 declares that certain types of interference with watercourses are statutory nuisances.

9.4 All the above categories of statutory nuisance are limited by the phrase 'to be prejudicial to health or a nuisance'. The phrase has been used in public health legislation for a long time, therefore the courts have laid down some guidance. It is not necessary for the local authority to prove both prejudice to health and nuisance in any single case – proof of *either* of the two characteristics is enough: see *Betts v Penge Urban District Council* [1942] 2 KB 154.

When Part IIA of the EPA comes into force, the statutory nuisance provisions will not apply to contaminated land which falls within Part IIA: see para 89 of Sch 22 to the Environment Act 1995.

'Prejudicial to health'

9.5 By s 79(7), 'prejudicial to health' means 'injurious or likely to cause injury to health'. Presumably, this originally meant prejudicial to *human* health, but it is arguable that the re-enactment of the statutory-nuisance provision in the wider-ranging 1990 Act includes the protection of flora and fauna. In *Coventry City Council v Cartwright* [1975] 1 WLR 845, the Court of Appeal held that, in order for an accumulation of rubbish to be prejudicial to health, it must be putrescible, with some likelihood of disease or that vermin would be attracted which could spread disease. The presence of inert material on the site was not a statutory nuisance notwithstanding that there was a chance that physical injury could be caused to people who walked on the site.

Unlike the 'nuisance' test (see below), mere interference with personal comfort does not amount to prejudice to health: see *National Coal Board v Thorne* [1976] 1 WLR 543 overruling *Betts v Penge Urban District Council* (above).

'Nuisance'

9.6 As to the word 'nuisance', it appears to be accepted that it reflects the common-law concept of nuisance: see Chapter 5 and *National Coal Board v Thorne* (above). As such, its parameters can be anticipated to some degree. In particular, it should be noted that land which causes damage to neighbours in its natural state may be a nuisance at common law and, thus, also for s 79: see *Leakey v National Trust* [1980] 1 All ER 17. It is not clear whether the usual common-law defences of prescriptive right and statutory authority apply in the context of statutory nuisance.

9.7 A nuisance can arise from interference with the personal comfort of a neighbour and it is not necessary that he suffers any injury: see *Bishop Auckland Local Board v Bishop Auckland Iron and Steel Company* (1882) 10 QBD 138. However, Lord Widgery CJ stated, in *Coventry City Council v Cartwright* (above), that the general purpose of the legislation was to 'stop accumulations which have a public health consequence'. This was interpreted by the Chelmsford Crown Court, in *Wivenhoe Port Limited v Colchester Borough Council* [1985] JPL 175, to mean that 'nuisance' could not include all public and private nuisances. In order to constitute statutory nuisance there must be a material effect on the personal well-being of those affected.

9.8 The common-law concept of nuisance is relevant to establishing whether there is an undue interference with the use and enjoyment of the land of another, but it does not follow that defences in tort or contract are available to the recipient of an abatement notice or order: see *A. Lambert Flat Management v Lomas* [1981] 2 All ER 280.

9.9 There is an important practical distinction between the 'prejudicial to health' and 'nuisance' tests, namely that a tenant may seek an abatement notice or order against his landlord under prejudice to health but not under nuisance. This is due to the fact that nuisance can only arise between neighbouring landowners – never landlord and tenant of the same premises: see *National Coal Board v Neath Borough Council* [1976] 2 All ER 478.

ACTION BY LOCAL AUTHORITIES: ABATEMENT NOTICES

9.10 If a person living in the local authority's area complains of a statutory nuisance, then the authority has a duty to take such steps as are reasonably practicable to investigate the complaint: s 79(1). The Secretary of State has default powers in the event that the authority fails to act: Sch 3, para 4.

9.11 If the local authority is satisfied that a statutory nuisance exists or is likely to occur or recur in its area, it shall serve an abatement notice requiring abatement of the nuisance or prohibiting or restricting its occurrence or recurrence and requiring the execution of such works, and the taking of such other steps, as may be necessary for any of these purposes: s 80(1). The phrase 'likely to occur' encompasses situations where the statutory nuisance might occur in the future and the requirement that the authority must be 'satisfied' indicates, on the usual principles of administrative law, that the authority must have acted reasonably in making that determination.

9.12 By s 80(2), an abatement notice will be served on 'the person responsible for the nuisance' but, if that person cannot be found or the nuisance has not yet occurred, then it will be served on the owner or occupier. If the nuisance arises from any defect 'of a structural nature' (undefined) then service is on the owner. There is nothing in the Act to preclude the service of multiple notices.

If more than one person is responsible for the nuisance, then one or all may be liable: s 81(1).

9.13 If any person on whom an abatement notice is served, without reasonable excuse, contravenes or fails to comply with any requirement or prohibition imposed by the notice, then he is guilty of a criminal offence: s 80(4). If the premises are industrial, trade or business premises then the maximum fine is £20,000, otherwise it is level 5 on the Standard Scale: s 80(5) and (6). It is essential to ensure, if a defendant is found guilty, that the penalty is not calculated on the higher scale (under s 80(6)) when the relevant premises are not 'industrial, trade or business' premises: see **9.29**. If the nuisance continues after the first conviction, then a further fine of one tenth of level 5 may be imposed for each day of that continuance.

9.14 It appears that before serving on an owner or occupier (in a non-structural case), the local authority has to make reasonable inquiries to identify the person responsible, but it does not have to make considerable efforts. In *Rhymney Iron Company v Gelligaer Urban District Council* [1917] 1 KB 589 a nuisance arose in relation to some drains. It was held that the authority did not have to trace the pollution through the drainage system to find the original polluter. The authority may need to produce evidence of its inquiries: see *R v Mead* [1898] 1 QB 110.

9.15 The authority also have power to enforce the abatement notice through the High Court (ie by injunctions) if it is of the opinion that fines imposed under the Act would afford an inadequate remedy: s 81(5).

Contents of the notice

9.16 There is no prescribed form for an abatement notice; however, it must notify the recipient of his rights of appeal: Sch 3, para 6. The notice must identify the nuisance and indicate required remedial measures. If there are self-evident remedies available, then it may simply prohibit the offensive activity and leave it to the recipient to decide how that is to be achieved. But if it requires that certain works be executed, or that the offensive activity be reduced to a particular acceptable level, then it must specify those works or that level with proper particularity: see *Millard v Wastall* [1898] 1 QB 342, *McGillivray v Stephenson* [1950] 1 All ER 942, and *Network Housing Association v Westminster City Council* [1994] EGCS 173.

The notice must specify the time or times within which its requirements are to be complied with: s 80(1). Reasonable time must be given for compliance: see *Bristol Corporation v Sinnott* [1918] 1 Ch 62. The prohibition of a recurrence is permanent (*R v Birmingham Justices ex parte Guppy* (1988) 152 JP 159) even after abatement and prosecution. Notices served under the Control of Pollution Act 1974 continue in force notwithstanding its repeal by the EPA: see *Aitken v South Hams District Council* (1994) *The Times*, 8 July, overruling *R v Folkestone Justices ex parte Kibble* (1993) *The Times*, 1 March.

Abatement by the local authority

9.17 Section 81(3) provides that where an abatement notice has not been complied with, the local authority may abate the nuisance and do whatever is necessary in execution of the notice. Section 81(4) then deals with the recovery of the costs of remediation:

> 'Any expenses reasonably incurred by a local authority in abating, or preventing the recurrence of, a statutory nuisance ... may be recovered by them from the person by whose act or default the nuisance was caused and, if that person is the owner of the premises, from any person who is for the time being the owner thereof; and the court may apportion the expenses between persons by whose acts or defaults the nuisance is caused in such manner as the court consider fair and reasonable.'

It is noticeable that unlike s 79, s 81(4) does not use the word 'sufferance', therefore, presumably, the range of persons chargeable under the subsection does not encompass all potential recipients of the notice: see discussion of s 79(7) at **9.27** and **9.28** below.

The reference to 'any person who is for the time being the owner thereof' apparently has the effect of making subsequent purchasers liable – thus demonstrating that proper inquiries need to be made at the time of purchase.

9.18 Section 81A provides that, where expenses are recoverable under s 81(4), the local authority may serve a notice on the owner which has the effect of making the expenses carry interest 'at such reasonable rate as the local authority may determine' and making the expenses and accrued interest a charge on the premises. Copies of the notice must be served on all persons who, to the knowledge of the authority, have any interest in the premises: s 81A(3). Any person served with a notice or copy thereof may appeal to the county court within 21 days of service: s 81A(6).

9.19 Section 81B provides that the local authority may, where the expenses are a charge on the premises under s 81A, declare that they are to be payable by instalments. Any sums recovered from an occupier may be deducted by him from his rent (with effect from 5 January 1994): s 81B(4). The liability of an occupier for repayment of expenses is geared to his liability for rent only: s 81B(5).

Powers of entry

9.20 Any person empowered by the local authority may on production (if so required) of his authorisation, enter any premises (ie the offensive property or neighbouring property) at any reasonable time for the purpose of ascertaining whether a statutory nuisance exists, the taking of samples, and, if necessary, carrying out appropriate remedial works: Sch 3. If the premises are residential, then the officer needs to give 24 hours' notice unless it is an emergency. The officer may, in certain prescribed circumstances, obtain a warrant from a justice of the peace to allow forced entry. It is a criminal offence to wilfully obstruct an officer using the power of entry.

The authority also has powers to require that persons interested in the land provide details of the owner, occupier and other persons interested in the land under s 16 of the Local Government (Miscellaneous Provisions) Act 1976.

Appeal against an abatement notice

9.21 The recipient of an abatement notice may appeal to a magistrates' court within 21 days of service of the notice: s 80(3). The grounds of appeal are set out in the Statutory Nuisance (Appeals) Regulations 1990 (SI 1990/2276) and the Statutory Nuisance (Appeals) (Amendment) Regulations 1990 (SI 1990/2483). They include (inter alia):

(a) that the abatement notice was not justified;
(b) that there was some error in the notice;
(c) that the requirements of the notice are unreasonable or unnecessary;
(d) that the best practicable means were used to counteract the effect of the nuisance from trade/business premises; or
(e) that the compliance period is too short.

The recipient is also entitled to argue that he is the wrong recipient or that other persons ought, also, to be served with enforcement notices – albeit that the court has no power to order that such further notices be served.

9.22 The court may uphold, quash or vary the notice. It may make orders as to who should carry out and pay for required works; it may appropriate any expenses between the parties – having regard to the terms of any tenancy. As to (e), *Nottingham Corporation v Newton* [1974] All ER 760 would suggest, although decided under earlier legislation, that the magistrates could allow an extended time period where, for example, the offending site is due to be cleared in the near future in any event. As to (d), the best practicable means of defence is discussed at **9.24** and **9.26** below. The notice will, except in certain prescribed cases, normally be suspended pending the outcome of the appeal.

ACTION BY THE PUBLIC: ABATEMENT ORDERS

9.23 Section 82 places a considerable power in the hands of the public and pressure groups. If a member of the public is aggrieved by an alleged nuisance then he is entitled to apply to the magistrates' court for an abatement order. An abatement order may be obtained against the local authority itself by a council-house tenant: see *Sandwell Metropolitan Borough Council v Bujok* [1990] 3 All ER 385. The court may, in addition to making the abatement order, fine the defendant up to level 5 on the Standard Scale.

A 'person aggrieved' must be more than a busybody – he must have some interest in the state of affairs arising from the nuisance. In *Birmingham City Council v McMahon* (1987) 51 JP 709, it was held that the occupier of a single flat could not be a person aggrieved in relation to disrepair and damp in the whole block of flats. Before instituting proceedings under s 82, the person aggrieved must give written notice of his intention to bring proceedings to the intended defendant: s 82(6). The minimum notice period is generally 21 days but with 3 days for noise complaints.

By s 82(4), the order may be made against 'the person responsible for the nuisance', but if that person cannot be found or the nuisance has not occurred, then against the owner or occupier. If the nuisance arises from any defect 'of a structural nature', then the order may be made against the owner.

If the person responsible, owner or occupier of the premises cannot be found or a person is convicted under s 82(8), then the court may order that the authority abate the nuisance: s 82(11) and (13). Breach of the order is a criminal offence: s 82(8). Both the 'reasonable excuse' and 'best practicable means' defences are available to the defendant: see **9.24–9.26**.

STATUTORY DEFENCES

9.24 There are two statutory defences to the offence of breach of an abatement notice (see **9.10–9.15**) or breach of an abatement order (see **9.23**) which are of relevance here, namely 'reasonable excuse' (ss 80(4) and 82(8)) and 'best practicable means' (ss 80(7) and 82(9)). The use of 'best practicable means' is also a ground of appeal against an abatement notice (see **9.21–9.22**).

9.25 'Reasonable excuse' is a defence if the defendant can show that a reasonable man would think that the excuse given was reasonable; it is an objective test. This defence has a very limited application. In *A. Lambert Flat Management v Lomas* [1981] 2 All ER 280, it was said that the defendant has a reasonable excuse when he has some special difficulty in complying with the notice, for example if he was out of the country at all relevant times. The court also decided that a defendant cannot use any of the grounds of appeal against an abatement notice as the basis for a reasonable excuse. A misunderstanding of the law does not amount to reasonable excuse: *R v Reid* [1973] 3 All ER 1020.

Lack of funds is not a reasonable excuse: *Saddleworth Urban District Council v Aggregate and Sand* (1970) 114 SJ 931. In *Wellingborough District Council v Gordon* [1991] JPL 874, Taylor LJ stated that reasonable excuse was limited to emergencies or circumstances beyond the defendant's control.

9.26 The 'best practicable means' defence is limited to abatement notices served on, or abatement orders made in respect of, industrial, trade or business premises and does not apply to smoke within s 79(1)(b) (unless it comes from a chimney) or to s 79(1)(c) or (h) at all: ss 80(8) and 82(10). As to 'industrial, trade or business premises' see **9.29**. In the case of an abatement order, the defence does not apply if a nuisance renders premises unfit for human habitation: s 82(10)(d).

The defendant will succeed if he shows that he used the 'best practicable means' to prevent or counteract the nuisance.

'Best practicable means' is extensively defined by s 79(9). 'Practicable' is defined as meaning reasonably practicable having regard, inter alia, to:

(a) local conditions and circumstances;
(b) the current state of technical knowledge; and
(c) the financial implications.

The 'means' to be employed include:

(a) the design, installation, maintenance, manner and periods of operation of plant and machinery; and
(b) the design, construction and maintenance of buildings and structures.

The test is to apply only so far as it is compatible with any legal duty, with safety and safe working conditions and with the exigencies of any emergency or unforeseeable circumstances. The burden of proving that best practicable means have been employed falls on the defendant: *Wivenhoe Port v Colchester Borough Council* [1985] JPL 175 and 396.

'PERSON RESPONSIBLE', 'OWNER' OR 'OCCUPIER'

9.27 As discussed above, ss 80(2) and 82(2) create various liabilities for 'persons responsible' and 'owners' and 'occupiers'. Section 79(7) defines a 'person responsible' as being a person 'to whose act, default or sufferance the nuisance is attributable'. Manifestly, this can catch landlords and mortgagees in possession provided that they have sufficient control to enable a court to take the view that the nuisance was due to their act, default or sufferance. To give one example, if a site is emitting an offensive smell due to landfill waste, and a mortgagee in possession has the ability to carry out works to stop that smell, then it is continuing by virtue of his default or sufferance. Owners or occupiers may fall within the definition if they allow a third party to cause the nuisance (eg by allowing the deposit of contaminated waste on their land) or if they continue a nuisance started by another. An original wrongdoer may remain liable even though he cannot, now, enter the land to stop the nuisance: see *Thompson v Gibson* (1841) 7 M&W 456.

Neither 'owner' nor 'occupier' are defined by the Act. However, there should be little difficulty in showing that a landlord or mortgagee in possession is an 'occupier' for the purposes of the Act. It is clear that freeholders are 'owners' and, arguably, so are sublandlords, lenders and receivers in possession. Certainly, the latter three categories would, if in possession, be 'occupiers'.

9.28 If the offensive property is occupied under a lease, then the authority is entitled to serve notices on either or both landlord and tenant – the authority is not obliged to look into the terms of the lease to decide how it apportions responsibility between them: *Wincanton Rural District Council v Parsons* [1905] 2 KB 34. Whilst the court may, under s 81(4) apportion the expenses of remedial works between recipients, it is not concerned to look behind leases or contracts in deciding the validity of the notice itself. The situation between landlord and tenant is, therefore, a private matter.

'Sufferance' includes allowing a nuisance to continue: see discussion of *Sedliegh-Denfield v O'Callaghan* in Chapter 5 (at **5.17**).

'INDUSTRIAL, TRADE OR BUSINESS PREMISES'

9.29 The phrase 'industrial, trade or business premises' is important in connection with ground (d) of s 79 (see **9.3**), the best practicable means of defence (see **9.24** and **9.26**) and whether the higher penalty under s 80(6) applies on breach of an abatement notice (see **9.13**). Section 79(7) of the EPA provides that 'industrial, trade or business' premises means premises used for any industrial, trade or business purposes or premises not so used on which matter is burnt in connection with any industrial, trade or business process – and premises are used for industrial purposes where they are used for the purposes of any treatment or process as well as where they are used for the purposes of manufacturing.

It is unclear whether this definition refers to the current use of a site or its historical or intended use. This raises a considerable problem when the relevant site is derelict. For example, is the site of a derelict factory 'industrial, trade or business premises' despite the fact that no activities related to those uses are now carried out there? Would it make any difference if the factory is currently being demolished to make way for a residential development? There is no clear answer; however, the initial words of the s 80(7) definition strongly suggest reference to the current use of the site. Reference to the historical use of the site is fraught with problems, not least being that of deciding how far back in time one needs to look. Is it the most recent past use or is one permitted to dredge through the history books, however old?

ANCILLARY COMPENSATION CLAIMS

9.30 Section 35 of the Powers of the Criminal Courts Act 1973 enables a court to order convicted persons to pay compensation (to a maximum of £5,000) to victims of their offences. In exercising its powers, the court has to have regard to the defendant's means – so far as they have been disclosed. Section 38 provides that the recovery of civil damages in respect of the same matter shall have regard to any compensation recovered under the Act. In *Herbert v Lambeth Borough Council* (1991) *The Times*, 27 November, an award of compensation was made following a conviction for breach of s 94 of the Public Health Act 1936. Woolf LJ stated that the principles of the decision would apply equally to statutory nuisances under the 1990 Act: see also *Botross v Hammersmith and Fulham London Borough Council* (1994) *The Times*, 7 November.

There is no reason why this procedure should not apply to private complainants under s 82 of the 1990 Act, thereby providing them with a relatively speedy route to compensation where the defendant has been convicted of breach of an abatement notice or abatement order.

Chapter 10

TOWN AND COUNTRY PLANNING

INTRODUCTION

10.1 This chapter will outline the town and country planning regime so far as is relevant to contaminated land, including that part of the regime which applies to hazardous substances on land. The technicalities of obtaining planning permission will be discussed in some detail in Chapter 11; this chapter will sketch the framework of the regime, introduce the basic concepts and indicate how the relevant authorities operate in practice.

Unless indicated otherwise, all references to statutory section numbers in this chapter refer to the Town and Country Planning Act 1990 (TCPA). The exception is the final section dealing with hazardous substances.

THE NEED FOR PLANNING PERMISSION

10.2 Planning permission is required for the 'development' of land: s 57. The concept is defined by s 55(1) as follows:

> '"development", means the carrying out of building, engineering, mining or other operations in, on, over or under land, or the making of any material change in the use of any buildings or other land.'

Section 55(3) provides that the deposit of refuse or waste materials on land involves a material change of use, notwithstanding that the land is comprised in a site already used for that purpose if:

- the superficial area of the deposit is extended; or
- the height of the deposit is extended and exceeds the level of the land adjoining the site.

The Town and Country Planning (General Permitted Development) Order 1995 (GDO) (SI 1995/418) provides an exception to s 55(3) for certain industrial uses pre-dating 1 July 1948: Sch 2, Part 8, Class D.

Statutory exceptions

10.3 Section 55(2) excludes certain works from the definition of development. They include:

(a) works for the maintenance, improvement or other alteration of any building which:
  (i) affect only the interior of the building; or
  (ii) do not materially affect the external appearance of the building and are not works making good war damage or begun after 5 December

1968 and providing additional underground space in the building;
(b) the use of any buildings or land within the curtilage of a dwelling-house for any purpose incidental to the enjoyment of the dwelling-house;
(c) the use of any land or any building occupied with the land for the purposes of agriculture or forestry (including afforestation); and
(d) the use of land in accordance with an order made by the Secretary of State.

10.4 Section 57 gives some cases of development where no planning permission is required and these include:

(a) resumption of the normal lawful use of land after expiry of a temporary permission;
(b) resumption of the normal use of land after a use permitted by the GDO; and
(c) use of the land for its lawful use after the issue of an enforcement notice.

The Use Classes Order

10.5 The Town and Country Planning (Use Classes) Order 1987 (UCO) (SI 1987/764) relaxes planning control by providing that, where premises are used for any purpose within a class specified in the order, then the use of those premises for any other use within the same class is not treated as development. Environmental matters play a part in the UCO. For example, the B1 class is as follows:

> 'Use for all or any of the following purposes:
> (a) as an office other than a use within class A2 (financial and professional services);
> (b) for research and development of products or processes; or,
> (c) for any industrial process
> being a use which can be carried out in any residential area without detriment to the amenity of that area by reason of noise, vibration, smell, fumes, smoke, soot, ash, dust or grit.'

Clearly, the question of whether or not premises fall within B1 will depend on environmental impact. Furthermore, it is possible that a use begins as a B1 use but, as time passes, smell or other pollution increases, thereby putting the use outside B1.

10.6 It is worth bearing in mind that developers often apply for, and local authorities sometimes grant, planning permission even though, in fact, it is not really needed. In *Newbury District Council v Secretary of State for the Environment* [1981] AC 578, planning permission was sought and granted in 1962 for use of aircraft hangars for the storage of synthetic rubber. The hangars had, previously, been used for storing agricultural materials and, before then, military hardware. The use had been, throughout, that of use as a 'repository' within class X of the Use Class Order 1950 and, as such, the 1962 permission was unnecessary. It is, therefore, necessary to check the UCO before deciding to apply for permission.

The General Permitted Development Order

10.7 The Town and Country Planning (General Permitted Development) Order 1995 has the effect of granting planning permission for a total of 76 classes of 'permitted development'. Development rights under the GDO are known as 'p.d. rights'. Special limitations apply to 'Article 1(5) land', namely land in special areas such as national parks, areas of outstanding natural beauty and conservation areas.

One occasionally finds that, as with the UCO, people often apply for permission when, in fact, the GDO gives p.d. rights. In *R v Swale Borough Council ex parte Royal Society for the Protection of Birds* [1991] JPL 39, planning permission was sought and granted for land reclamation in the Medway Estuary using the spoil from dredging the Medway Channel. Although the case was decided on other grounds, Simon Brown J (as he then was) opined that the development was permitted under the 1988 version of the GDO. As such, one should check the GDO before making an application.

AUTHORITIES

10.8 Local planning authorities are district councils, unitary councils and county councils or, in the case of land inside Greater London, London borough councils or the common council of the City of London. An urban development corporation is the local planning authority (LPA) for its own area. In a national park, the LPA is the relevant national park committee. An LPA can compromise a joint planning board covering the areas of two or more councils. The planning powers of the county councils are limited to 'county matters'. These are matters relating to the disposal and treatment of waste materials (see Annex 7 of Planning Policy Guidance Note (PPG) 23) and specified classes of minerals development.

In practice, the functions of LPA's are usually exercised by the committees of elected representatives. Certain matters, usually of minor importance, are delegated to specific officers of the authority under s 101 of the Local Government Act 1972. An LPA cannot delegate the discharge of its functions to a 'committee' of one member: *R v Secretary of State for the Environment ex parte Hillingdon London Borough Council* [1986] JPL 363.

DEVELOPMENT PLANS

10.9 Outside Greater London and the former metropolitan counties the 'development plan' can generally be said to consist of:

(a) the structure plan together with (where applicable) the Secretary of State's notice of approval;
(b) alterations to that plan;
(c) The local plan or (where relevant) the waste local plan or minerals local plan and the resolution of approval or adoption; and
(d) alterations to the local plan, waste local plan or minerals local plan.

The county council prepares the structure plan for its area. The plan formulates, in broad strategic terms, the county council's policy for the development and use of land in its area. It does not deal with particular parcels of land. The district planning authorities usually prepare local plans. A local plan is a map and a written statement showing the LPA's detailed proposals for the development and other use of land in that part of its area. These proposals must generally conform to the structure plan and their function should be to give detail to the strategic proposals in the structure plan. A local plan will not (subject to specified exceptions) contain any policies in respect of development which involves the depositing of refuse or waste materials: s 36(1). Such policies will be contained in the waste local plan or minerals local plan: s 38(2).

10.10 The Greater London Council and the metropolitan county councils were abolished from 1 April 1986. Within London and the metropolitan areas, any structure plan, local plan or old style development plan which was in force at the date of abolition continued to be effective until superseded by a unitary development plan (UDP), the new forward planning mechanism in these areas. The UDP is an amalgam of structure plan and local plan, with two parts. Part I is a strategic plan containing general policies and proposals in respect of development and other use of land in the authority's area. Part II contains more detailed policies and proposals, including policies which would be in the waste local plan elsewhere. Upon adoption or approval, the UDP supersedes all earlier plans, and so it becomes the development plan for the area. As from April 1996, a number of county councils and district or borough councils will be abolished and replaced by unitary authorities. These unitary authorities will be the LPAs for their areas. Pre-existing structure plans and local plans will continue in force until new unitary development plans for the new areas are produced: see the Local Government Changes for England Regulations (SI 1994/867).

Section 54A

10.11 On 25 September 1991 s 54A of the 1990 Act came into force. It has added considerable weight to the development plan in planning decisions and states:

> 'Where, in making any determination under the planning Acts, regard is to be had to the development plan, the determination shall be made in accordance with the plan unless material considerations indicate otherwise.'

The prudent developer must, therefore, pay particular attention to any relevant development plan policies before making any formal overture to the LPA.

The development plan and contaminated land

10.12 Development plans have a significant role to play in environmental protection and enhancement. All development plans must include policies in respect of the conservation of the natural beauty and amenity of land and the improvement of the physical environment: ss 12(3A), 31(3) and 36(3). However, as a matter of law, structure plans and Part I of UDPs must have regard to the resources likely to be available: ss 12(6)(c) and 31(6)(c). PPG 12 adds that *all* development plans (ie including local plans) should be realistic in this respect and emphasises the need to reuse redundant, derelict or underused sites in preference to greenfield sites: para 5.20. PPG 12 indicates that plan-making authorities should carry out an environmental appraisal as part of the plan preparation process: paras 5.52 and 6.25. The PPG then notes that policies must weigh and reconcile economic and social priorities (including the need for economic growth) against environmental priorities: paras 6.4 and 6.5.

10.13 More recently, PPG 23 has advised that structure plans should include strategic land use policies on the location of potentially polluting developments and on the location of sensitive developments in the vicinity of existing polluting development: para 2.15. It advises that local plans should include an appropriate combination of site-specific policies for potentially polluting development, and the criteria against which applications for such development may be determined, and policies for development on, and remediation of, existing contaminated and derelict land: para 2.17. Local plan policies should aim to keep housing and other developments sensitive to pollution apart from polluting or potentially polluting uses and take account of the environmental consequences of former land uses: para 2.18.

PPG 23 advises that, in preparing their development plans, planning authorities need to take into account the possible effects on health and the environment of contaminated land and that development plans should provide an opportunity to set out policies for the reclamation and possible use of contaminated land: para 4 of Annex 10.

10.14 Paragraph 5 of Annex 10 to PPG 23 states:

> 'Local plans and Part II of Unitary Development Plans should include detailed criteria which will be applied in determining planning applications for development on land which is known to be, or may be, contaminated. They may also set out any site-specific proposals for land use, where contamination is known, or the site history suggests a risk of contamination, so that they may be readily identifiable to landowners and prospective purchasers or developers.'

There are a number of points arising from that paragraph.

First, in the references to site history, there is a strong suggestion that the Department of the Environment (DoE) is revisiting the debate about registers of land subject to contaminative uses which centred on s 143 of the Environmental Protection Act 1990: see Chapter 15. Some local planning authorities will read this advice as allowing them to make statements about the history of particular sites in their local plans, thereby impacting on the value of such sites.

Secondly, the reference to 'detailed criteria' might tempt some LPAs to provide over-detailed control. One example of the problems which can arise comes from proposed policy ME8 from the draft of the Bristol City Local

Plan (as placed on deposit in November 1993) which reads:

> 'For all development proposals on land which is contaminated the City Council will require appropriate remedial measures to be taken to minimise any risk of pollution within the site and in the surrounding area, as a condition on the granting of planning permission.'

The proposed policy looks innocuous and appears to comply with advice from central government; however, the reasoned justification which follows creates a problem:

> 'For all sites which are contaminated or *suspected of being contaminated* the City Council will require developers to carry out a *detailed site survey* to determine the nature, extent and level of contaminants present both in the soil and underlying geology *before the application is determined.* The City Council will expect to be appraised of the scope of the study and the sampling measures used. In cases where contamination is shown to exist a *detailed* scheme showing the appropriate remedial measures required will need to be agreed *before* planning permission is granted' (emphasis added).

As will be seen from the discussion of PPG 23 in Chapter 11, advice from central government (which repeated earlier advice on this point) is to the effect that it is not necessary to incur the costs of a detailed site investigation where there is only a suspicion that the site is contaminated. This proposed policy therefore places an unnecessary burden on the developer.

10.15 A draft policy to similar effect can be found in the Consultation Draft Northavon Local Plan published in July 1994:

> 'N32 Proposals for development of land which has been identified as actually or potentially contaminated will be expected to be accompanied by a detailed site survey and analysis to determine the type and amount of hazardous substances present. The Council will require measures to remove the contamination or render it harmless as a condition of granting planning permission.'

Given the weight attributed to development plans by s 54A, the message for landowners, prospective purchasers and developers is clear – they ought to participate in the development plan process to ensure that their sites are not subject to onerous and unnecessary policy control. All development plans are subject to public consultation procedures and, as such, owners and developers of land have the opportunity to make objections to proposed policies. Clearly, such objections need to be made and pursued where proposed development plan policies will have an adverse effect on land.

GUIDANCE FROM CENTRAL GOVERNMENT

10.16 Guidance from central government is in two main forms, circulars and PPGs. DoE circulars contain advice on new legislation, technical matters and, until superseded in this regard by the PPG series (see below), policy advice. The DoE from time to time issue PPGs, intended to provide guidance on general and specific aspects of planning policy. Additionally, a series of

Mineral Planning Guidance Notes (MPGs) deal with the control of mineral development. PPGs and MPGs are intended to provide concise and practical guidance on planning policies in a clearer and more accessible form than Departmental circulars. The DoE's long-term objective is to replace policy statements in its circulars by statements in PPGs.

The DoE also publishes Regional Planning Guidance Notes (RPGs) which set out very broad-based long-term regional planning policies. For example, RPG 10 was published in July 1994 giving guidance for the South-West to the year 2011. It covers Avon, Cornwall, Devon, Dorset, Gloucestershire, Somerset and Wiltshire.

10.17 When looking at a planning problem one should always refer to these policy statements in addition to the text of the principal Act and relevant regulations. The wording of circulars and guidance notes are very often the main source of submissions at planning appeals and inquiries. Of particular importance in the context of contaminated land are PPG 14 *Development on Unstable Land,* PPG 23 *Planning and Pollution Control* and MPG 7 *The Reclamation of Mineral Works*. General guidance on the role and operation of the planning system can be obtained from PPG 1 and PPG 12; and from Circular 11/95 on planning conditions.

The guidance discussed in this section can be purchased from HMSO.

THE GRANT OF PLANNING PERMISSION

10.18 Planning permission may be granted either in outline or in detail. A detailed planning permission will normally be complete and, subject to compliance with its conditions, will be ready for immediate implementation. An outline planning permission establishes that the proposed development is acceptable in principle but further consent has to be obtained in relation to the details – called 'reserved matters'.

Planning conditions

10.19 Section 70(1) provides that an LPA may grant permission subject to such conditions as they think fit. Section 72 then provides that such conditions may be imposed in respect of land outside the application site if that land is in the 'control' of the applicants. In *George Wimpey v New Forest District Council* [1978] JPL 314, the High Court held that 'control' does not mean having an estate or interest in the land but only such right as was required to implement the condition. Despite the wide wording of the statute, the LPA does not enjoy an unfettered discretion. In particular, its power is constrained by the principle of 'reasonableness': see **3.12–3.14**. A condition may also fail if it is uncertain in effect or absurd. Some conditions are imposed by statute, for example the duration of a planning permission (s 92).

The use of conditions in the context of contaminated land is discussed in **11.21–11.24**.

Planning obligations

10.20 The phrase 'planning obligation' arose as a result of amendments made to s 106 of the TCPA 1990 by the Planning and Compensation Act 1991. Planning obligations now take the place of what were commonly called 'section 52 agreements'. A planning obligation may be created either by a planning agreement or a unilateral undertaking. As its name implies, a planning agreement is a document entered into between the landowner and the LPA. By contrast, a unilateral undertaking is a deed executed by the landowner only and delivered to the LPA. Unilateral undertakings have a role to play in planning appeals where the landowner is seeking to persuade the Secretary of State to grant permission on the basis of, among other things, the offer of some 'planning gain'. That gain can be guaranteed by the execution of a unilateral undertaking. Planning obligations run with the land provided they have been registered as local land charges.

The use of planning obligations in the context of contaminated land is discussed in 11.21–11.24.

Aftercare conditions

10.21 Section 72(5) of and Sch 5 to the 1990 Act provide that LPAs may impose aftercare conditions when granting, revoking or modifying planning permissions relating to the deposit of refuse or waste materials. It is an offence for any persons in control of such land to fail to comply with a notice requiring him to comply with such a condition: see **10.31**.

ENVIRONMENTAL ASSESSMENT

10.22 Planning applications for certain projects (including 'deemed applications' under the Town and Country Planning General Regulations 1992 (SI 1992/1492)) must be accompanied by 'environmental statements' and, before the planning authority determines such an application, it must consult with various public bodies and engage in an 'environmental assessment'.

The requirement for environmental assessments derives from EC Directive 85/337 and has been transposed into UK law by the Town and Country Planning (Assessment of Environmental Effects) Regulations 1988 (SI 1988/1199) as amended by the Town and Country Planning (Assessment of Environmental Effects) (Amendment) Regulations 1994 (SI 1994/677). Advice from the DoE on the topic is contained in Circular 15/88 and the DoE publication *Environmental Assessment – A Guide to the Procedures* (HMSO). The DoE has also published a consultation paper entitled *Environmental Assessment and Planning: Consultation Draft Guide on the Preparation of Environmental Statements* (September 1994).

The need for environmental assessment

10.23 Projects listed in Sch 1 to the 1988 regulations must be preceded by an environmental assessment. These are major projects such as oil refineries, power stations, motorways and waste-disposal installations for the treatment or landfill of dangerous or toxic substances.

For projects in Sch 2, environmental assessment is required if the proposed development is likely to have significant effects on the environment by virtue of its nature, size or location. Obviously, the question of what amounts to 'significant effects' can be the cause of considerable debate. In the first instance, the LPA has a wide 'fact and degree' discretion in determining whether an environmental assessment is necessary: see *R v Swale Borough Council ex parte Royal Society for the Protection of Birds* [1991] JPL 39. In cases of dispute, the matter can be referred to the Secretary of State for his determination. In Circular 15/88 the Secretary of State advises that Sch 2 projects which may need environmental assessment are of three main classes:

- major projects which are of more than local importance;
- occasionally, smaller projects at particularly sensitive or vulnerable locations; and
- in a small number of cases, projects with unusually complex and potentially adverse environmental effects.

Content of environmental statement

10.24 If the developer volunteers or is required to provide an environmental statement, then it must contain the 'specified information' prescribed by Sch 3 to the 1988 regulations. It must describe the development proposed and the direct and indirect likely significant effects of the proposals on the environment by reference to a range of targets including human beings, fauna, soil, water and air. No mention is made of effects on flora, but human 'cultural heritage' is listed. The environmental statement will, once submitted, become a public document and it is subject to mandatory publicity requirements. In practice, the environmental statement for a very large or complex project will normally be preceded by a 'scoping report' which outlines the matters to be covered by the statement. The scoping report will be sent to the planning authority and the statutory consultees for their comments.

Proposed amendments to the 1985 Directive

10.25 In May 1994, the European Commission issued proposals to amend Directive 85/337, the main ones being to add two new stages to the procedure. First, a 'screening' stage, whereby competent authorities are positively required to consider whether or not environmental assessments should be required in cases which meet specified criteria. This procedure is designed to overcome the problem that projects which, on the face of it, should have been subject to environmental assessment have slipped through the net because the competent authorities have simply failed to ask whether environmental assessment is necessary. Secondly, a 'scoping' stage, whereby the competent authority defines, in advance of the environmental statement, the information to be provided by the developer. This is designed to overcome the problems caused by many statements containing only minimal information.

The draft Directive also proposes changes to consultation and decision-making procedures: see ENDS Report 232.

THE ENFORCEMENT OF PLANNING CONTROLS

10.26 If 'development' (as defined by s 57) is carried out without planning permission or a condition or limitation forming part of a permission is not complied with, then there is a breach of planning control which may be subject to enforcement action by the local planning authority.

Time-limits

10.27 If the breach of planning control consists of operational development without planning permission, then the LPA cannot take enforcement action later than 4 years following substantial completion of the operations: with all other breaches (save in relation to dwelling-houses), the time-limit is 10 years from the date of the breach: see s 171B. Subject to some exceptions outside the scope of this work, a breach of planning control does not give rise to any immediate criminal liability.

Enforcement notices

10.28 The remedies for breach of control are, with the exception of the planning injunction, administrative remedies – and the mainstay remedy is the 'enforcement notice'. The LPA may issue an enforcement notice requiring the owner or occupier of the relevant land to remedy an apparent breach of planning control if it is expedient to do so, provided that the time-limits in s 171B are observed. If the recipient or any person with an interest in the land wishes to dispute the notice, then he is entitled to appeal to the Secretary of State for the Environment: s 174. The notice is suspended pending the outcome of such appeal: s 175(4).

Subject to the determination of any appeal against it, failure to comply with an enforcement notice is a criminal offence: s 179. It is a defence for the defendant to show that he did everything he could be expected to do to secure compliance with the notice: s 179(3). The authority may also enter the land, carry out the measures required by the enforcement notice and recover the costs of doing so from the owner of the land: s 178. In *R v Collet and Others* [1994] 2 All ER 372 the House of Lords held that knowledge of the existence of an enforcement notice is not essential for criminal liability under what is now s 179. Those occupying land were under an obligation to advise themselves of the planning status of the land. Section 179(7) now provides that it is a defence for the defendant to show that he was not aware of the existence of the notice if: (a) he has not been served with a copy of the notice; and (b) it was not kept on the appropriate statutory register.

10.29 If there has been the breach of a condition or limitation forming part of an extant planning permission then, as an alternative to an enforcement notice, the planning authority may serve a 'breach of condition notice' (see **10.31**) under s 187A.

Stop notices

10.30 Neither enforcement notices nor breach of condition notices are effective immediately: both must prescribe a period for compliance of not less than 28 days. This can cause problems where an immediate remedy is necessary. Thus, under s 183, the authority may serve a 'stop notice' with the enforcement notice or afterwards: but a stop notice may not be served when the enforcement notice has taken effect. The stop notice will normally provide that an ongoing breach of planning control cease within 3 days but, if there are special reasons, the authority may give the stop notice immediate effect. There is no appeal against a stop notice.

Breach of condition notice

10.31 Section 187A of the 1990 Act came into force on 27 July 1992 and provides that, where a condition of a planning permission is not complied with, the LPA may serve a breach of condition notice (BCN) on the developer or any person having control of the relevant land, requiring compliance with the planning conditions specified in the BCN. The expression 'conditions' in s 187A is deemed to include 'limitations': s 187A(13)(a). The concept of limitations is much debated within planning circles and there is considerable doubt both in terms of defining the concept and its aegis. A limitation is, very broadly, part of the description of the development itself separate from the conditions in the permission and it can be gleaned from the words of the permission and accompanying documents. Potentially, therefore, s 187A has a very wide ambit, albeit an uncertain one.

Failure to comply with a BCN is a criminal offence: s 187A(9). A person charged under s 187A(9) may raise the statutory defences that (a) he took all reasonable measures to comply with the BCN, or (b) that he no longer had control of the land when the notice was served.

There is no direct right of appeal against a BCN; however, the recipient could achieve the same effect by making an application for removal or modification of the condition to the LPA (under s 73) and then appealing to the Secretary of State for the Environment against any refusal. If the defendant believes that the BCN is ultra vires, then he could challenge it by judicial review (see **3.10–3.18**) or might be able to raise the issue as a defence in any criminal proceedings: see *Wandsworth Borough Council v Winder* [1985] AC 461 and *R v Reading Crown Court ex parte Hutchinson* [1988] 1 All ER 333; but see also *R v Wicks* (1995) *The Times*, 19 April.

Planning injunctions

10.32 In extremis, the planning authority may apply to the county court or High Court for a planning injunction preventing a current or anticipated breach of planning control: s 187B.

SECTION 215 NOTICES

10.33 Section 215 provides that, if it appears to the LPA that the amenity of a part of their area, or of an adjoining area, is adversely affected by the condition of land in their area, they may serve on the owner and occupier of the land a notice requiring that steps be taken to remedy the condition of the land within a period (of not less than 28 days) specified in the notice. The scope of s 215, in relation to pollution, is limited because it is based on adverse affect on 'amenity'. It would cover unpleasant odours and visual damage (eg dead vegetation, etc), but is somewhat narrow in the current context.

10.34 The recipient of the notice or any other person holding an interest in the land is entitled to appeal against the notice on a number of grounds, including that the damage to amenity is attributable to and as such results in the ordinary course of events from the carrying on of operations or a use of the land which is not in breach of planning control: s 217. It remains to be seen exactly how the phrase 'ordinary course of events' will be applied by the courts, but it is difficult to see how s 217 could justify pollution caused by the implementation of planning permission in a sloppy fashion, for example where a permitted factory emits unpleasant smells or allows the leakage of deleterious chemicals.

Where such an appeal is brought, the relevant notice is suspended pending the final determination or withdrawal of the appeal: s 217(4). There is a further right of appeal to the Crown Court: s 218.

10.35 Failure to comply with an effective notice is a criminal offence punishable by a level 3 fine and a daily penalty for each day that the breach continues after first conviction: s 216. Section 216(3) and (4) provide that where a past owner or occupier of the site is being prosecuted, then he may bring his successor owner or occupier before the court. The past owner or occupier may then avoid conviction if he can prove that the failure to comply with the notice was attributable (in whole or in part) to his successor and that he took all reasonable steps to ensure compliance with the notice: s 216(5).

If the steps required by the notice have not been taken in the specified time, then the local planning authority may enter the land, execute those steps and recover the reasonable costs of doing so from the current owner: s 219(1). If another person was responsible for the condition of the land, then the owner has a right of recourse to him: s 219(2).

PLANNING APPEALS

10.36 If an LPA refuses to grant a planning permission, or grants a permission subject to conditions or limitations which are unacceptable to the applicant, or delays making a determination beyond the period prescribed by the GDO, then the applicant may appeal to the Secretary of State for the Environment: s 78. The effect of such an appeal is that the Secretary of State will determine the application as though it were made to him in the first instance: s 79(1).

In the majority of cases, the Secretary of State will appoint an Inspector to hear and determine the appeal; however, the Secretary of State will make the final determination himself on major and important appeals. One of the material considerations which will be taken into account by the Secretary of State or his Inspector on an appeal, will be the Secretary of State's own guidance as published by the DoE in circulars and PPGs. That is one of the reasons why such guidance carries considerable weight with LPAs – hence the emphasis given to guidance from the DoE in this and the following chapter.

HAZARDOUS SUBSTANCES

Hazardous substances consent

10.37 Consent must be obtained from the Hazardous Substances Authority (HSA) to the presence on, over or under land of any hazardous substance in excess of the 'controlled quantity': s 4 of the Planning (Hazardous Substances) Act 1990 (P(HS)A). The Planning (Hazardous Substances) Regulations 1992 (SI 1992/656) list 71 substances which fall within the P(HS)A and set their 'controlled quantities'. By s 4(2) of the P(HS)A, the 'controlled quantity' is gauged by reference to the aggregate quantity of the substance:

(a) on, over or under the land;
(b) on, over or under other land which is within 500 m of it and controlled by the same person; or
(c) in or on a structure controlled by the same person any part of which is within 500 m of it.

The Act does not, as such, define 'hazardous waste'.

10.38 There are a number of exceptions to the need for hazardous substances consent (HSC) including:

(a) where the substance is present (but not unloaded) temporarily whilst in transit;
(b) certain substances contained in aerosols;
(c) substances in 'service pipes' and exempt pipelines; and
(d) substances present for no more than 14 days as a result of an emergency.

Controlled waste (see Chapter 8) and radioactive waste (as defined by the Radioactive Substances Act 1960) are excluded.

It is important to note that there is an overlap between the P(HS)A, the Notification of Installations Hazardous to Health Regulations 1982 (SI 1982/1357) and the Control of Major Accident Hazard Regulations 1984 (SI 1984/1902) in terms of controlled substances; therefore a site covered by either the 1982 or 1984 regulations may still need HSC.

Application procedure

10.39 The application procedure follows, very broadly, the procedure used for obtaining planning permission and includes mandatory publicity and notification requirements. The HSA must also engage in extensive consultation with bodies including the National Rivers Authority, English Nature, and the Health and Safety Executive. As to the latter, DoE Circular 11/92 advises, at para B15, that the HSA should not override the Executive's advice to refuse an application 'without the most careful consideration'.

There are provisions where HSC is deemed to be consented if the hazardous substance was on the land before 1 June 1992 and a claim was made before 1 December 1992: s 11. An HSC may be granted subject to conditions: s 9. Section 10 allows the LPA to require the partial or whole completion of a development permitted by a planning permission in certain specified cases as a condition of the grant of an HSC. Applications may be made to vary HSC conditions.

Revocation and variation

10.40 An HSC will normally run with the land, in the same way as planning permission, but this depends on any transfer being of the *whole* site and not a part of it. If there is a change of control of a part of the land to which the HSC relates, then the HSC will be revoked automatically unless an application for its continuation has been made: s 17. Section 17 does not apply where the whole site is transferred.

10.41 Section 14(1) of the P(HS)A gives the HSA a general power to revoke or modify an HSC, but this power is subject to the payment of compensation to any person whose interest in the land, or enjoyment of it, is depreciated or who carries out works pursuant to the order: s 16.

Section 14(2) allows the revocation or modification of an HSC without compensation if the order is confirmed by the Secretary of State (see s 15) and if:

(a) there has been a material change of use of the land to which the HSC relates;
(b) planning permission for the material change of use of the land to which the HSC relates has been granted and commenced;
(c) where the HSC is for one hazardous substance only and it has not been kept at the land in controlled quantities for at least 5 years; and
(d) where the HSC is for more than one hazardous substance, where none of the substances have been kept on the land for at least 5 years.

The criminal offences

10.42 If any person allows or knowingly causes a hazardous substance to be present on, over or under land at or exceeding the controlled quantity without an HSC or in an amount exceeding the maximum quantity allowed by an HSC, then that person *and* any person in control of the land are both guilty of an offence: s 23(1), (2) and (3). It is also an offence to breach a condition in an HSC, but only the person in control may be prosecuted: s 23(3)(b). The offences are triable either way: s 23(4).

There are statutory defences in s 23(5), (6) and (7) namely:

(a) that the defendant took all reasonable precautions and exercised due diligence to avoid commission of the offence;
(b) that the offence could have been avoided only by taking action which amounted to breach of statutory duty; and
(c) that the defendant did not know and had no reason to believe –
  (i) for a s 23(2)(a)(i) offence, that the substance was present or that it was present in a quantity equal to or exceeding controlled quantities;
  (ii) for a s 23(2)(a)(ii) offence, that the HSC maximum controlled quantity was exceeded; and
  (iii) for a s 23(2)(b) offence, that there was failure to comply with an HSC condition.

Contravention notices

10.43 The HSA may, where there has been a breach of hazardous substances control, issue a hazardous substances contravention notice and serve copies on the person in control of the land and all persons having an interest in it: s 24. The contravention-notice regime is closely modelled on the enforcement-notice procedures under the TCPA: see **10.28** and **10.29**. The Planning (Hazardous Substances) Regulations 1992 provide for rights of appeal against the notice on grounds similar to those applying to enforcement notices under the TCPA. The Secretary of State can, on appeal, quash or vary the contravention notice.

Failure to comply with an extant contravention notice is a criminal offence: see Sch 4, Part 5 to the 1992 regulations.

Injunctions

10.44 Under s 26AA of the P(HS)A, the HSA can apply for an injunction in respect of an actual or apprehended breach of hazardous substances control without exhausting other remedies.

Chapter 11

CONTAMINATED LAND AND PLANNING PERMISSION

INTRODUCTION

11.1 Almost all significant commercial development requires planning permission and this chapter will explore the particular problems which arise when permission is sought for development of land that may be contaminated. Town and country planning is about striking balances, and economic arguments have often to be weighed against environmental considerations. In his address to the Joint Planning Conference in September 1993, the Junior Minister for Local Government, Inner Cities, Construction and Planning, Mr Tony Baldry MP, stated:

> 'There will always be differences of view – between those who want to build and those who want things left as they are. What is needed is a mechanism for balancing economic growth with environmental protection.
>
> If we are honest – this is a monstrous slander! We want economic growth and an improved environment. We want new jobs, new schools, and new hospitals that growth brings. But we also demand, that such growth is sustainable – that it is not at the expense of our children.'

In deciding the location of a new hospital, the planner may be forced to decide between using a greenfield site – and suffering the ecological impacts that may bring – or placing it on a contaminated site – and accepting the costs and risks *that* may bring. The UK is a small country with a very limited range of options, almost all of them involving difficult choices.

11.2 The Government has stated its keenness to see derelict and contaminated land reclaimed and reused. The Department of the Environment (DoE) consultation paper *Paying for Our Past* indicated that:

> 'The role of the planning system in encouraging the investigation, reporting, assessment and remediation of land as part of the process of ensuring its beneficial use or re-use will continue.'

Paragraph 4.1 of Planning Policy Guidance Note (PPG) 23 (see **11.4** below) states that:

> 'The principle of sustainable development means that, where practicable, brownfield sites, including those affected by contamination, should be recycled into new uses and the pressures thereby reduced for greenfield sites to be converted to urban, industrial or commercial uses. Such recycling can also provide an opportunity to deal with the threats posed by contamination to health or the environment.'

THE DUPLICATION OF CONTROLS

11.3 One of the problems which arises in allocating responsibility in this way is that, of course, other regulatory regimes have been specifically created to deal with various aspects of pollution, and there is a risk of planners duplicating the regulation imposed by others. This is undesirable because it can create conflicts between regimes and involves the burden of unnecessary bureaucracy.

Planning Policy Guidance Note 23

11.4 Advice on these issues was produced by the DoE in July 1994 in PPG 23, *Planning and Pollution Controls*, together with specific advice on the planning regime and contaminated land.

Paragraph 1.2 of PPG 23 states:

> 'The planning and pollution control systems are separate but complementary in that both are designed to protect the environment from the potential harm caused by development and operations, but with different objectives.'

Paragraph 1.33 adds that:

> 'The role of the planning system focuses on whether the development itself is an acceptable use of the land rather than the control of the processes or substances themselves.'

Paragraph 1.3 states:

> 'The planning system should not be operated so as to duplicate controls which are the statutory responsibility of other bodies (including local authorities in their non-planning functions). Planning controls, except where they are applied in the context of hazardous substances consents, are not an appropriate means of regulating the detailed characteristics of potentially polluting activities.'

Paragraph 1.11 of the PPG sees the complementary role of the planning system as regulating:

- the location of development and the control of operations in order to avoid or minimise adverse effects on the use of land and on the environment; and
- what happens after any development or use of land, so that land (and water resources) are restored to such a condition as to be capable of the agreed after-use, once any development or use of land has ceased.

Paragraph 1.32 states that the planning system should also control other development in proximity to potential sources of pollution.

The *Gateshead* case

11.5 The problem of overlapping regimes came before the Court of Appeal in *Gateshead Metropolitan Borough Council v Secretary of State for the Environment and Others* [1995] JPL 432, where an application for planning permission to construct a clinical waste incinerator in Gateshead was refused on grounds which included the uncertain environmental effects of, and paucity of information on, the proposals, in terms of the possible release of noxious substances and potential long-term health risks. The Secretary of State allowed the subsequent appeal, but did so against the recommendation of an Inspector and an Assessor that permission be refused on environmental grounds. The Secretary of State accepted that it was necessary to have regard to environmental considerations but stated that it was not the role of the planning system to duplicate controls under the Environmental Protection Act 1990 (EPA). The EPA provided stringent controls over the potential emissions. The Secretary of State's decision was upheld at first instance by Mr Jeremy Sullivan QC sitting as a Deputy Judge of the High Court: [1994] JPL 255.

The Deputy Judge's decision was upheld by the Court of Appeal on 12 May 1994. Glidewell LJ stated that the environmental impact of the proposed discharges was a material planning consideration, and accepted the Deputy Judge's earlier statement that the existence of a stringent regime under the EPA for preventing or mitigating that impact was also a material planning consideration. The appellants argued that, once planning permission had been granted, there was no real prospect of Her Majesty's Inspectorate of Pollution (HMIP) refusing to authorise the operation of the plant. It was argued that HMIP were unlikely to do more than ensure that the best available techniques not entailing excessive costs be used. The Court of Appeal rejected that argument: Glidewell LJ could see no reason why HMIP could not refuse authorisation if they so determined. The concerns which led the Inspector and the Assessor to recommend refusal were matters which could be properly decided by HMIP.

11.6 It is important to stress that nothing in the *Gateshead* decision diminishes the materiality of environmental matters as planning considerations. The Deputy Judge added that his decision was not carte blanche for applicants to say 'leave it all to the EPA'. Applicants must still satisfy the local planning authority (LPA) that environmental concerns have been met in full or that they can be later regulated by another regime. If, as in the *Gateshead* case, there is considerable uncertainty as to the environmental impact of a project, then the applicant will have to satisfy the LPA that the relevant regulating body has the ability to veto the project in the event that the environmental impact is unacceptable. It is notable that, in the *Gateshead* case, HMIP had given evidence to the Inspector to the effect that the appellant's proposals provided a 'valid starting point for authorisation procedures'. It is also notable that the application was for outline permission; therefore the LPA still enjoyed control over reserved matters.

Consultation between authorities

11.7 PPG 23 stresses that its aim is to encourage close consultation and prevent unnecessary duplication and conflict of interest between planning and pollution-control authorities: para 1.4. In many circumstances, an LPA must, as a matter of law, consult with a variety of other bodies before coming to a planning decision including:

- the National Rivers Authority (NRA) where, inter alia, the development consists of storage of mineral oil and derivatives, or use of the land for waste, refuse, sewage or sludge, or the reuse of land so used or near land which is being so used;
- the waste regulation authority, the Ministry of Agriculture, Fisheries and Food and the Forestry Commission on the use and aftercare of landfill sites; and
- the Health and Safety Executive on proposals to redevelop contaminated land.

THE 'SUITABLE FOR USE' APPROACH

11.8 It is important to note that LPAs should not try to persuade developers to engage in the overly rigorous remediation of contaminated sites. Paragraph 4.2 of PPG 23 states:

> 'The Government's policy towards tackling the UK burden of land contamination is that the works, if any, required to be undertaken for any contaminated site should deal with any unacceptable risks to health or the environment, taking into account its actual or intended use.'

The paragraph goes on to add what appears to be a qualification:

> 'This "suitable for use" approach does not preclude an owner, occupier or developer from undertaking earlier or more thorough action if he wishes to do so. Nor does it preclude a regulatory body from requiring very thorough remedial works where the circumstances justify it.'

It remains to be seen what regulatory bodies will make of the phrase 'very thorough remedial works'; however, it is submitted that that phrase should not be taken out of context and used as justification for remedial requirements which go beyond the 'suitable for use' approach.

The 'suitable for use' approach is explored in more detail in Chapter 13.

SITE INVESTIGATIONS

11.9 One of the key questions to be answered for any would-be developer of a contaminated site is whether comprehensive site investigations need to be carried out before planning permission is granted. This issue bears upon cash-flow and viability calculations. If the LPA requires that, before it considers the principle of development on the site, the applicant must spend a considerable sum for a detailed site investigation, then the project might not be viable from the outset. A better alternative for the developer is for the planning authority to grant outline planning permission, subject to conditions requiring the site investigation before the reserved matters are determined. The grant of outline permission will establish the broad principle of development on the site and thus create the value which will underpin the cost of the site investigation.

Guidance can be obtained from ministerial advice, but that advice differs between applications for sites which are or may be contaminated and applications on or near closed or operating landfill sites. It is necessary to look at these situations separately.

Non-landfill sites

11.10 Guidance on applications which do not involve landfill sites or nearby land can be obtained from PPG 23 and Circular 11/95. The PPG and the Circular distinguish between situations where there is only a suspicion that a site may be contaminated and those where contamination is known or strongly suspected.

Paragraph 9 of Annex 10 to the PPG states:

> '... where there is *only a suspicion* that the site might be contaminated, or where the evidence suggests that there may be only slight contamination, planning permission may be granted but conditions should be attached to make it clear that the development will not be permitted to start until a site investigation and assessment has been carried out and that the development itself will need to incorporate all the measures shown in the assessment to be necessary' (emphasis added).

Paragraph 75 of the Circular is expressed in similar terms.

The approach to be adopted where there is a *strong suspicion* of contaminated or *actual knowledge* of its existence is more rigorous. Paragraph 8 of the Annex indicates that:

> '*However, if it is known or strongly suspected that the site is contaminated to an extent which would adversely affect the proposed development or infringe statutory requirements*, an investigation of the hazards by the developer and proposals for any necessary remedial measures required to deal with the hazards will normally be required before the application can be determined by the local planning authority' (emphasis added).

Paragraph 74 of the Circular is expressed in almost identical terms. Paragraph 76 adds:

> 'Conditions might also be imposed requiring the developer to draw to the attention of the planning authority the presence of significant unsuspected contamination encountered during the redevelopment.'

A decision letter in an appeal by Barrett Northampton Ltd against a refusal of planning permission for residential development of a contaminated site at Irthlingborough gives some useful guidance on the issue of planning conditions in this context: see [1995] JPL 759. There the site was a former tannery site and was known to be contaminated. The Inspector stated:

> 'Paragraph 9 of Annex 10 to PPG 23 advises that where it is known that a site is contaminated to an extent which would adversely affect the proposed development, proposals for remedial measures will normally be required before the application is determined. To my mind, in the event of a grant of permission this advice clearly points to the need for specific and detailed

conditions relating to remediation. I have reached this conclusion in the light of the advice in the following paragraph of the PPG, which allows for a non-specific approach to conditions in cases where there is only a suspicion that the site might be contaminated, or where the evidence suggests that there may be only slight contamination.'

In determining whether or not the site fell within paragraph 8 of Annex 10, the Inspector took into account evidence based on ICRCL Guidance Note 59/83 (see **13.5**) and found that the site fell within para 8 because the relevant 'threshold' concentrations were exceeded:

> '[ICRCL 59/83] sets out "trigger" concentrations of contaminants; for those found on the appeal land, "threshold" levels provide an indication of the requirement for remedial action. The survey work done on the site of the former tannery found frequent concentrations above the ICRCL 59/83 threshold levels for arsenic and chromium, and less widespread instances of lead contamination and cadmium. A concentration of toluene extract (not dealt with by ICRCL 59/83) was also found in the vicinity of the site of the tannery boilerhouse ... In this case, there is no dispute over the levels of contamination, which, taking into account the ICRCL 59/83 thresholds, appear to me to be substantial at least with regard to certain contaminants on various parts of the appeal site.'

Circular 11/95 also provides some model conditions which may be referred to by local planning authorities in drafting conditions in respect of the development of contaminated land.

> 'Contaminated land
>
> 56. Development shall not begin until a scheme to deal with contamination of the site has been submitted to and approved in writing by the local planning authority.
>
> 57. The above scheme shall include an investigation and assessment to identify the extent of contamination and the measures to be taken to avoid risk to the [public/buildings/environment] when the site is developed.
>
> 58. Development shall not commence until the measures approved in the scheme have been implemented.
>
> *Where investigation/remedial proposals carried out/agreed before planning permission granted.*
>
> Soil decontamination
>
> 59. Before the development hereby permitted commences on the site, a soil survey of the site shall be undertaken and the results provided to the local planning authority. The survey shall be taken at such points and to such depth as the local planning authority may stipulate. A scheme for decontamination of the site shall be submitted and approved by the local planning authority in writing and the scheme as approved shall be fully implemented and completed before any [residential] unit hereby permitted is first occupied.
>
> *For use where soil contamination is known or suspected.'*

Model conditions 58 and 59 will not be appropriate to many developments and need to be used with care. For example, the contamination may affect only a small part of a large development site and it might be unreasonable to hold up the entire development until this issue is resolved.

11.11 Where an application for outline planning permission is not accompanied by sufficient information on which to reach a decision, LPAs may request supplementary information under Art 3 of the Town and Country Planning (General Development Procedure) Order 1995 (SI 1995/419). However, PPG 23 makes it clear that applications need not be delayed pending a detailed investigation by the developer to establish the nature and extent of the contamination: Annex 10, para 6.

Landfill sites

11.12 There are particular problems associated with the development of landfill sites and nearby land, especially where the construction of buildings is proposed. The principal hazards are flammable and explosive landfill gases. Also, geotechnical conditions may be unsuitable for construction purposes, substances in the soil may attack building materials, and there may be problems associated with odour, site drainage, leachate production and surface run-off. General advice is contained in DoE Circular 17/89.

11.13 Paragraph 19 of Circular 17/89 states:

> 'Where the redevelopment of a closed landfill site or of nearby land is involved, *even the possibility of difficulties from migrating gas would be a material planning consideration.* This might result in permission being refused, or warrant imposing appropriate conditions to manage any gas from the closed landfill, or to ensure that suitable precautions are taken with construction methods. Where the presence of gas has been discovered, or it is suspected that it may be present during the development of a site, investigations should be carried out to determine the source of the gas and apply any remedial measures to prevent it causing a hazard either during the course of development or during subsequent use of the site' (emphasis added).

Thus for the development of a closed landfill site or of nearby land, even the possibility of difficulties from migrating gas is a material planning consideration. The radius of 250 m around a closed landfill site is normally taken as the area of concern, but there is no hard and fast rule. Contamination can actually migrate for many hundreds of metres, especially where the subsoil is porous and contains conduits such as natural fissures or old mineshafts. Therefore, it is sensible to combine an environmental assessment with surveys of ground and geological conditions.

11.14 Paragraph 21 of the circular goes on to state that:

> 'Local authorities should exercise due caution in granting permission for development or redevelopment on or near landfill sites. Permission should not be granted unless reliable arrangements can be made to overcome the danger of migrating gas.'

Under Art 10 of the Town and Country Planning (General Development Procedure) Order 1995, LPAs are required to consult with waste-regulation authorities on proposals to develop sites within 250 m of land which:

(a) is or has, at any time in the 30 years before the relevant application, been used for the deposit of refuse or waste; and
(b) has been notified to the LPA by the waste-regulation authority for purposes of this provision.

Paragraph 10 of the Circular 17/89 then adds that 250 m is a minimum and LPAs ought to consult more widely on suspect sites.

11.15 It is difficult to see how an LPA could properly discharge its functions if it did not require the application to supply survey results at the time of the application. In 1992 a developer had costs awarded against him after a failed planning appeal in respect of proposed residential development on a closed landfill site. The Secretary of State made the award because the developer had pursued the appeal without providing adequate survey material on ground conditions and potential gas production. The decision is also notable for the reliance placed on ICRCL 17/78 (see below): DoE ref. APP/D1520/A/130168.

11.16 Circular 17/89 refers developers and planners to the Interdepartmental Committee on the Redevelopment of Contaminated Land (ICRCL) and its Guidance Note 17/78 on the development and after-use of landfill sites. The ICRCL notes take a particularly strong line on the redevelopment of landfill sites or nearby land. They suggest that, because of the real physical hazards, such sites are generally unsuitable for 'hard' forms of development such as buildings or structures unless appropriate remedial treatment is carried out before construction begins:

> 'Where there is landfill gas there should be a very strong presumption against development which involves construction of traditional types of housing (ie houses with gardens) which would require particularly stringent precautions such as the provision of high rates of sub-floor and cavity ventilation and the installation of gas detectors and alarms ... Blocks of flats, offices, supermarkets, etc may be less difficult to protect because the sub-floor ventilation requirements can be provided, for example, by raising the building clear of the ground – a form of construction which is unsuitable for traditional types of housing development – and other means of meeting ventilation requirements which are more capable of being subjected to quality assured management' (p 1).

The position is more difficult where the landfill site is on land outside the control of the applicant. One possibility is the requirement that warning systems may be installed in built development. The problem is, of course, that such alarms need to be maintained during the years of the gas emission and this could pose problems. This is especially so where, as ICRCL 17/78 indicates, the development is sold off in small parcels of residential development. Likewise, a multiplicity of ownership's could cause problems for associated protective measures such as ventilation requirements. If, therefore, the developer is considering residential development, he may have to consider blocks of flats or some form of elaborate management scheme over traditional housing. There is no legal difficulty in setting up a management company to oversee a housing estate; however, there is a considerable marketing problem because British housebuyers prefer the autonomy created by clear freehold.

11.17 One can also foresee that the LPA will be reluctant to allow full 'permitted development' rights to obtain after the construction of development. The GDO permits the carrying out of certain minor development without the need for planning permission, for example the construction of sheds and garages on residential sites. However, the LPA has power to restrict such permitted development rights by issuing a direction under Art 4 of the GDO. Such direction would be justified where the LPA is aware of a landfill-gas problem and is seeking to limit the construction of buildings within which gas may collect and hard surfaces which may trap gas or cause it to migrate.

CONDITIONS AND OBLIGATIONS

11.18 The problem for a potential developer is clear: detailed site investigations are invariably very expensive and even a cursory one can run into thousands of pounds. He may be unlikely to want to invest heavily unless the principle of development has been accepted through the grant of, at least, outline planning permission.

Negative conditions and obligations

11.19 One option, where there is a suspicion of contamination affecting the development site, may be to seek permission subject to a negative planning condition (a 'Grampian' condition) or a planning obligation under s 106 of the TCPA to the effect that no development shall take place unless and until the site investigation has been carried out: see the model conditions discussed at **11.10**. In *Jones v Secretary of State for Wales and Ogwr Borough Council* [1990] JPL 907, the Court of Appeal took the view that it would be unlawful for an LPA to grant planning permission subject to a 'Grampian' condition unless there is a reasonable prospect that the impediment to implementation will be removed within a reasonable time.

The *Jones* case was overruled and the law restated by the House of Lords in *British Railways Board v Secretary of State* [1994] JPL 32. There it was held that the question of whether or not a negative condition can be complied with is a matter for the developer and that, normally, the LPA is entitled to impose a negative condition even if it believes that there is no reasonable prospect that the condition will be discharged. This is a considerable relaxation in relation to conditions requiring that site investigations and remedial measures be approved by the planning authority before the proposed development is started. Prior to the *British Railways* case, one objection to such a condition was that there could be no certainty that it would be discharged – but normally that objection cannot now be raised as a matter of law.

However, the Secretary of State takes the view that:

> '... the House of Lords established that the mere fact that a desirable condition, worded in a negative form, appears to have no reasonable prospects of fulfilment does not mean that planning permission must necessarily be refused as a matter of law. However, the judgment leaves open the possibility for the Secretary of State to maintain as a matter of policy that there should be at least reasonable prospects of the action in question being performed within the time-limit imposed by the permission' (footnote 3 to Circular 11/95).

Accordingly, para 40 of the Circular states:

> 'It is the policy of the Secretaries of State that such a condition should only be imposed on a planning permission if there are at least reasonable prospects of the action in question being performed within the time-limit imposed by the permission.'

This advice needs to be borne in mind when considering the negative conditions referred to by PPG 23 and Circular 11/95 in this context: see **11.10** above.

A contaminated land condition will often carry a degree of uncertainty over its fulfilment because, often, the nature of any remedial works will not become clear until an environmental assessment has been carried out. In many cases the remedial works may prove to be completely impracticable; in others the works may have to be carried out on land other than the application site and outside the control of the applicant. As such, there could be cases where non-specific conditions may be acceptable for the purposes of PPG 23, but are unacceptable for Circular 11/95. In that latter case, the applicant will be faced with either carrying out site investigations before the grant of the permission or a 'refusal based on failure to comply with 11/95'.

Outline permissions

11.20 Paragraph 3.21 of PPG 23 states that:

> 'In granting outline permission, authorities should ensure that the development permitted is restricted so it cannot take place in a form which would lead to significantly different environmental effects from those considered at the planning application stage. In cases where the environmental effects, possibility of pollution and public health risk are not of sufficient significance to affect the principle of whether the development should be permitted, planning authorities may attach conditions to the outline planning permission requiring particular environmental matters to be considered in more detail at a later stage.'

The PPG adds that, once an outline permission has been granted, it would be unreasonable to attach conditions to the detailed approvals of environmental matters that have the effect of nullifying the outline permission: para 3.21. However, PPG 23 notes that an outline planning permission is not usually appropriate for a development where the risk of pollution is significant: para 3.20.

Avoidance of duplication

11.21 It is important to stress that planning conditions and planning obligations should be directed towards planning objectives only. PPG 23 notes that the increasing scope and effectiveness of pollution-control regulations is reducing the need to use planning conditions or planning obligations to control pollution aspects of a development that is subject to other statutory regulation: para 3.23. This therefore puts the planning authority in the position of having to examine whether the environmental impacts of a development will be subject to effective control by a pollution-control authority. If the answer is in the affirmative, then the planning authority should not impose planning conditions which duplicate the roles of other bodies. Paragraph 3.24 of PPG 23 states:

> 'Planning authorities should not therefore seek to control through planning measures matters which are the proper concern of the pollution control authority, except where planning interests can be clearly distinguished.'

This advice is repeated, in the context of planning conditions, by para 22 of Circular 11/95, which states that a condition which duplicates the effect of other controls will be ultra vires because it is unreasonable.

THE PROTECTION OF GROUNDWATER

11.22 Groundwaters provide 35% of the public water supply in England and Wales and comprise an important resource. One likely area of concern in the future is the problems associated with the contamination or potential for contamination of groundwater. For example, leachate produced by the infiltration of rainfall through contaminated soil or landfill waste can be a source of pollution to ground or surface waters.

The NRA has shown itself to be very active in bringing prosecutions for pollution and has the technological ability to trace contaminating materials and chemicals back to source. Prospective developers should be aware of the NRA document *Policy and Practice for the Protection of Groundwater* (1992), which sets out the NRA's strategy in relation to the planning process including objecting to proposals for the redevelopment of contaminated sites that would adversely affect groundwater. In 1993 the NRA started a 3-year programme to map the more important extraction points and create 'source protection zones' around them. The NRA has already published 'groundwater vulnerability maps' at a scale of 1:1,000,000 and is producing more detailed versions at a scale of 1:100,000. The DoE has stated that these maps are material considerations to be taken into account in the drawing-up of development plans. PPG 12, on development plans, states:

> '6.19 Particular attention should be paid to the protection of groundwater resources which are susceptible to a wide range of threats arising from land-use policies. Once groundwater has been contaminated, it is difficult if not

> impossible to rehabilitate it. Changes in land use may also affect the availability of groundwater resources by restricting [discharge] or diverting flows. The National Rivers Authority is preparing a series of maps to identify those areas of particular concern: these will become available over the next three years. They should be taken into account in drawing up development plans.'

11.23 The protection of groundwater is also a material consideration in individual planning decisions. In November 1991 the Secretary of State refused an appeal by Leigh Environmental Ltd for a waste treatment centre at Doncaster, saying 'the location of the proposal above an important unprotected acquifer so near the surface provides sufficient reason to outweigh the argument concerning need and the other benefits associated with the proposal': DoE ref APP/F4410/A/89/126733.

Furthermore, it is necessary to have regard to EC policy on this topic: notably the Directive on groundwater protection discussed at **7.22**. At the end of February 1992, the Council of Ministers resolved to prepare an action programme for the protection of groundwaters, noting that the conservation of groundwater is a 'major political, economic and environmental imperative' which can be assured only by the implementation of the principle of sustainable development.

THE PRECAUTIONARY PRINCIPLE

11.24 It will be seen from Chapter 13 that environmental assessment is far from an exact science, which could mean that the LPA is left in doubt as to the probable effects of a proposal. If that situation arises, then it is necessary to ask how the authority ought to cope with that doubt. Should it err on the side of the developer? Unfortunately for the developer, the 'precautionary principle' indicates that the authority ought to err on the side of caution.

The Maastricht Treaty formally introduced the 'precautionary principle' as a fundamental tenet of European law (see **4.44**), and it has been an objective of UK environmental policy since the 1990 White Paper *This Common Inheritance* (para 1.18). Whilst specifically directed towards pollution-control systems, there can be little doubt that para 1.12 of PPG 23 makes the Government's general position clear:

> 'The Environment White Paper stressed that the Government's approach to pollution issues is a precautionary one. Where there are significant risks of damage to the environment, pollution controls will take into account the need to prevent or limit harm, even where scientific knowledge is not conclusive.'

However, this statement does not amount to a legal requirement. In *R v Secretary of State for Trade and Industry, ex parte Duddridge*, (1994) *The Independent*, 4 October, the High Court held that the observation of the precautionary principle is not a legal requirement, but it will carry considerable weight as an expression of government policy.

Chapter 12

NATURE CONSERVATION

INTRODUCTION

12.1 Contaminated land is often disused or derelict land which has not been subject to activity for years or decades. It is axiomatic that if man does not use land, then nature will. Flora may colonise disused land and create habitats for a variety of fauna. This can present a problem for the would-be developer, because, if the flora or fauna has rarity value, then the local planning authority (LPA) may be reluctant to grant planning permission for its redevelopment. The developer might find that the prospect of redevelopment has triggered the designation of the site for nature conservation purposes. Indeed this point has been picked up by at least one LPA in its draft local plan. One of the changes proposed to the Deposit Bristol City Local Plan in October 1994 states:

> 'Of particular concern is that some sites, after years of dereliction, have developed considerable nature conservation or amenity value. The ecological value of derelict sites and their amenity value as open spaces will be given a high priority in deciding their future uses.'

Also, para 15 of Planning Policy Guidance Note (PPG) 9, *Nature Conservation* (Department of the Environment, 1994) observes that 'the skilled adaptation of derelict areas can provide extended habitats'.

This philosophy can even lead to the compulsory acquisition of a derelict site as a local or national nature reserve. This has occurred on a number of occasions, most recently in Bristol where a derelict railway embankment which contained no significant rare species of either flora or fauna was expropriated from the owner for such a purpose: see **12.24**. The compulsory purchase order was upheld by the Secretary of State for the Environment, after a public inquiry, and sets a worrying precedent for potential developers of derelict land.

12.2 There are some express statutory duties in relation to nature conservation. Sections 61A, 61B and 61C of the Land Drainage Act 1991 (as amended) place statutory duties on internal drainage boards, ministers, the National Rivers Authority, local authorities and the Nature Conservancy Council (Countryside Council in Wales) to further nature-conservation interests. The Town and Country Planning Act 1990 (TCPA) requires that unitary development plans, structure plans and local plans include policies in respect of natural beauty, the amenity of land and improvement of the physical environment: ss 12(3A), 31(3) and 36(3).

This chapter will involve a limited examination of nature conservation law and policy as it applies to the disused or derelict site. Unfortunately, it leads to the conclusion that the prudent purchaser or developer will not only engage in the environmental assessments described in Chapter 13, but will also, before becoming too committed, have to engage the services of an ecological expert to assess any nature-conservation constraints.

Nature conservation involves the protection and study of special flora and fauna and the topic can be roughly divided into (1) the protection of individual flora and fauna, and (2) the protection of their habitats.

THE PROTECTION OF FLORA AND FAUNA

12.3 This section will examine the statutory protection given to certain individual flora and fauna. It is important, at the outset, to deal with a fundamental misconception about statutory protection – the idea that, because a particular species is protected, it must necessarily follow that it has rarity value or is, in some way, special. Statutory protection does not always follow ecological importance. The Protection of Badgers Act 1992 protects badgers, yet there are over 250,000 of them in England and Wales. Badgers are protected because some people would otherwise subject them to 'sport' which involves cruelty and inhumane treatment. However, the presence of a protected species will be a material consideration in a planning determination: see para 47 of PPG 9.

The main statute giving rise to the protection of individual flora and fauna is the Wildlife and Countryside Act 1981 (WCA). The Act provides broad protection for all classes of wild bird and selective protection of other fauna; it can be a minefield for the unwary. Taking the case of wild plants, some are protected yet some are not. In reality only the expert or the exceptionally well educated will be able to distinguish between the two classes – but criminal liability can turn on the distinction.

Wild birds

12.4 By s 1 of the WCA it is an offence to:

- kill, injure or take any wild bird;
- take, damage or destroy a nest of any wild bird whilst it is in use or being built; or
- take or destroy an egg of any wild bird.

Game birds are generally excluded from the protection of the Act and the protection in relation to wildfowl is restricted.

12.5 Section 4 provides some statutory defences. Of special note is that it is a defence if the action was an incidental result of a lawful operation and could not reasonably have been avoided. This is a defence which is unclear in its ambit. In many cases the defence could turn on intent. For example, it is a lawful operation to fell a tree which is outside a conservation area and is not protected by a tree preservation order (see **12.10–12.12** below), but what if the feller's only motive was to get rid of a bird nesting in the tree? In that case the nest damage is not an 'incidental result' of the operation because it was its main objective. But let us suppose that the feller is toppling the tree because it stands in the way of a major development project. In that case the defence might apply because the object was to remove the tree (not the bird) and the nest damage was an unavoidable incident of that operation. Section 4 also allows the removal of or damage to nests if necessary for the protection of public health and safety.

A site clearance based on a general desire to rid the site of its ecological value (perhaps as a preliminary to a planning application) would probably fall outside the defence. The use of the phrase 'reasonably avoided' clearly indicates that the person alleged to be in breach of the Act must show that he acted reasonably having regard to all material considerations, including, presumably, financial ones. In cases of doubt, the prudent developer should carry out any site clearance outside nesting season.

Protected animals

12.6 Section 9 of the WCA provides protection for the animals listed in Sch 5 to the Act. These include all bats, reptiles and amphibians and specified rare fauna. It is an offence to kill, injure, take or have in one's possession (whether alive or dead) any such animal. Sections 10 and 16 provide defences and exemptions similar to those applying to wild birds, but there is a special procedure requiring consent from English Nature in respect of bats. It is also an offence to damage or obstruct the place of shelter or protection of a scheduled wild animal. It is not clear exactly what is meant by 'place of shelter or protection'. Clearly, it covers nests, dens and the like, but it is arguable that the phrase also includes land in the immediate vicinity. This is an extended meaning, but, if correct, one which causes numerous problems for the developer.

The WCA is now supplemented by the Conservation (Natural Habitats & c) Regulations 1994 (SI 1994/2716), which make it an offence to deliberately kill, injure, take or disturb certain listed animals or to destroy their nesting places or breeding sites.

Badgers

12.7 Badgers enjoy special protection under the Protection of Badgers Act 1992. It is an offence to wilfully kill, injure or take (or attempt to do so) a badger or to (intentionally or by recklessness) interfere with a badger's sett: ss 1 and 3. A 'badger's sett' is defined as any structure or place which displays signs indicating current use by a badger: s 14. The phrase 'current use' should not need any further definition; however, English Nature appears to take the view that it covers seasonal use at some time during the course of a year, even if it is not used for the whole year, whereas the Ministry of Agriculture, Fisheries and Food (MAFF) takes it to mean showing signs of being in use or with signs of recent usage: see *Problems with Badgers?* 3rd revised edn (RSPCA, 1994). It is difficult to see how English Nature reaches its conclusion, because 'current' plainly means 'current' and not 'intermittent' or 'seasonal'. The MAFF approach appears to be the correct one.

If it is necessary to interfere with or destroy an active badger sett, then an application for consent may be made to English Nature: s 10. If consent is given, then it may be on the basis that a new sett be created nearby. There are a number of exceptions and statutory defences under ss 6, 7 and 8 – including where a badger is unavoidably killed or injured as an incidental result of a lawful action: s 6.

Protected plants

12.8 By s 13 of the WCA, it is an offence for any person other than the owner or occupier or other authorised person to intentionally uproot any wild plant. To put this into perspective, it is illegal to pick a dandelion whilst walking in the open countryside! Schedule 8 lists certain wild plants which may not be uprooted by any person. Subject to specified defences, it is an offence to do so. The Conservation (Natural Habitats & c) Regulations also make it an offence deliberately to pick, collect, cut, uproot or otherwise destroy certain listed plants.

Protected trees

12.9 Trees are an essential element in the environmental make-up of the planet in numerous ways and also enhance the visual amenity of this country. Yet, surprisingly, they are not subject to any nationwide blanket protection. The nearest approximation is the Forestry Act 1967, which makes it a criminal offence to fell growing trees without a felling licence from the Forestry Commissioners. However, there are numerous qualifications and exceptions to this general limitation. A landowner may fell up to 5 m^3 of trees on land occupied by himself or his tenant in any quarter, or engage in permitted felling for thinning purposes. Also, there are a series of general exceptions including felling to abate or prevent a nuisance or to comply with a planning permission.

Tree preservation orders

12.10 If a local authority has designated an area as a conservation area then, under s 211 of the TCPA, it is an offence for any person to lop, top, or cut down any tree within the conservation area without giving 6 weeks' notice to the LPA and then either waiting until that period expires or obtaining the authority's consent. If the authority object, then they can use that 6-week period to issue a tree preservation order (TPO).

If the local authority is concerned to protect trees, then the usual procedure is by way of a TPO under ss 197–201 of the TCPA and the Town and Country Planning (Tree Preservation Order) Regulations 1989. A TPO may cover specific trees, groups of trees, woodlands or all trees within a specified area: s 198(1). The LPA must consider that it is expedient in the interests of amenity to make the TPO. Whilst authorities often make area TPOs, this approach is often criticised because it is implicit that, in making the TPO, the authority has not considered the amenity value of individual trees or groups of trees on the site. The Secretary of State has a rarely used residual power to make TPOs under s 202.

12.11 The TPO procedure is designed to give immediate effect to 'provisional TPOs' only, with non-provisional TPOs taking effect only when they are confirmed. In practice, TPOs almost always start out as provisional TPOs, because some landowners would quickly fell trees before the confirmation of a non-provisional TPO.

Copies of the draft order will be served on all owners and occupiers of affected land and it will be placed on deposit for public inspection. Objections may be made to the TPO within a period specified in the order (but not being less than 28 days) and the LPA will either confirm, modify or withdraw the TPO after consideration of those objections. If the TPO is a non-provisional TPO, then it will come into effect only when confirmed. If it is in the provisional form, then it will lapse after 6 months unless it has been confirmed in that period.

In practice, the authority will often have to act with great speed (eg where felling contractors are on site) and will often issue an area-based provisional TPO immediately and then follow up with a conventional TPO (during the 6-month protection period) based on a tree survey specifying trees or groups of trees. The modification of an area order to a woodland order (as opposed to issuing a fresh TPO) may be unlawful in certain circumstances: see *Evans v Waverley Borough Council* [1995] EGCS 132.

12.12 Breach of a TPO is a summary offence subject to a level 4 fine, but if any person cuts down, uproots or wilfully destroys a TPO tree or wilfully damages, tops or lops it in such a way as to be likely to destroy it, then this is an offence triable either way with an unlimited fine in the Crown Court or a maximum of £20,000 in the magistrates' court: s 210 of the TCPA. The offences are based on strict liability, therefore it is no defence that the defendant was unaware of the TPO: see *Maidstone Borough Council v Mortimer* [1980] 3 All ER 552.

There are a number of statutory defences to the breach of a TPO including:

- where the tree was dead, dying or in a dangerous condition;
- if done to abate a nuisance;
- if done to comply with any Act of Parliament; and
- if done pursuant to a felling licence issued by the Forestry Commission.

Section 206 requires the replacement of any tree removed in breach of a TPO or because it was dead, dying or in a dangerous condition. If any person proposes any operation which could breach a TPO, then he must obtain consent from the LPA. A conditional TPO consent or a refusal may give rise to compensation unless the authority certifies that the refusal was in the interests of good forestry or that the tree or trees had outstanding or special amenity value: s 203. Such a refusal or conditional consent also gives the applicant a right of appeal to the Secretary of State.

THE PROTECTION OF HABITATS

12.13 Areas or sites of particular ecological value may be protected by a variety of statutory or policy designations which could create problems for the developer. Such designations (eg sites of special scientific interest (SSSI)) may arise under statute, and designation will have day-to-day implications as well as restricting development opportunities. Other designations may be policy designations (eg local plan policies) which come into play only when development proposals are mooted.

There are currently some 3,500 SSSIs in England together with 43 special protection areas, 49 Ramsar (see **12.31**) sites and over 240 national nature reserves: see *Action for the Countryside* (Department of the Environment, 1992). The task of understanding this topic is complicated not only by the multiplicity of possible designations but also because many sites carry more than a single designation. Thus an SSSI could also be subject to a 'nature conservation order' under the WCA and could be allocated as a 'site of wildlife importance' in the local plan.

Sites of special scientific interest

12.14 Section 28 of the WCA provides that where English Nature (or the Countryside Council for Wales) is of the opinion that any area of land is of special interest by reason of its flora, fauna or geological or physiographical features, it shall be its duty (as a precursor to designation as an SSSI) to notify that fact:

(a) to the LPA in whose area the land is situated;
(b) to every owner and occupier of any of that land; and
(c) to the Secretary of State.

The notification will give a list of operations which English Nature consider are potentially damaging to the features of interest on the site.

Section 28 appears to be limited to the *current* interest of the site as opposed to its future or potential interest, albeit that English Nature appears to see potential value as a secondary criterion for selection: see their publication *Guidelines for the Selection of Biological SSSIs*. Furthermore, the section appears to be limited to the interest of the site as a site and does not include designation where the site does not have individual interest but has some form of 'group value', for example as part of a 'wildlife corridor'. It would appear that the designation can include adjacent areas which would not, by themselves, qualify under s 28 provided that they form part of the same 'environment': see *Sweet v Secretary of State for the Environment* [1989] 1 JEL 245.

Section 28 is cast in terms of giving English Nature a *duty* to designate, therefore, in theory, it can be compelled to designate by an interested person or pressure group when the necessary evidence has been provided. Whether that theory would translate into practice remains to be seen.

12.15 The notification by English Nature will specify a period of not less than 3 months for the making of representations and objections. The proposed SSSI designation must be withdrawn or confirmed within 9 months of original notification: s 28(4A). There is no statutory appeal procedure.

Once designation is complete, it is an offence for any owner or occupier to carry out a 'potentially damaging operation' on the SSSI without reasonable excuse, unless the procedure prescribed by s 28(5) of the 1981 Act is observed. Somebody who is not an owner or occupier (eg a contractor) cannot be prosecuted under s 28: *Southern Water Authority v Nature Conservancy Council* [1992] 3 All ER 481. Under s 28(5) and (6)(c), the owner or occupier may write to English Nature asking for consent to the operations and, if consent is not forthcoming within 4 months, may carry out those operations in any event after the expiry of the 4-month period. The 4-month period will allow English Nature to seek a nature conservation order (see **12.16–12.19**) or the compulsory acquisition of the site as a nature reserve (see **12.20–12.25**).

This procedure need not be followed in an emergency (provided that English Nature is notified as soon as practicable after commencement of the operations) or if planning permission has been granted for the operations (not permission granted by a general or special development order).

Nature conservation orders

12.16 Under s 29 of the 1981 Act, the Secretary of State may, where it is expedient to do so, make a nature conservation order (NCO) in respect of land which is of special interest by reason of its flora, fauna, or geological or physiographical features:

(a) for the purpose of securing the survival in Great Britain of any kind of animal or plant or of complying with an international obligation; or
(b) where, in the opinion of the Secretary of State, the land is of national importance, for the purpose of conserving its flora, fauna or geological or physiographical features.

The order will specify those operations which cannot be carried out on the land unless certain procedures in s 29 are observed.

An NCO comes into effect as soon as it is made, but must be notified to the owner, occupier and local authority within 28 days: Sch 11. Those affected may object within 28 days and, if those objections are not withdrawn, the Secretary of State for the Environment will have to decide whether or not to confirm it – usually after a public inquiry.

12.17 A specified operation which appears to the Secretary of State to be likely to destroy or damage the features that caused the NCO to be made may only be carried out if the owner or occupier of the land has given notice to English Nature and:

(a) English Nature gives its consent;
(b) the operation is carried out in accordance with a management agreement under s 16 of the National Parks and Access to Countryside Act 1949 (NPACA) or s 15 of the Countryside Act 1968; or
(c) 3 months have expired since the giving of the notice.

The 3-month period in (c) may be extended to 12 months if English Nature, before that period expires, offers to enter into an agreement under s 16 of the NPACA or s 15 of the Countryside Act 1968.

If English Nature makes a compulsory purchase order during this extended period, then it will be further extended until English Nature enter the land after the order or until it is withdrawn or confirmation is refused.

12.18 Breach of the NCO in the way described in **12.17** (ie breach of s 29, without reasonable excuse, is a criminal offence triable either way with a possible unlimited fine in the Crown Court: s 29(8). As with a conventional SSSI, it is a reasonable excuse if the operation was authorised by a planning permission or carried out in an emergency: s 29(9). Upon a conviction under s 29, the court may order the person convicted to restore the land to its condition prior to the operations: s 31. If he fails to do so, then English Nature may restore the land and recover the costs of doing so from him: s 31(6).

12.19 The owner or occupier of land subject to an NCO is entitled to compensation: s 30 and Wildlife and Countryside (Claims for Compensation under Section 30) Regulations 1982 (SI 1982/1346). With agricultural land, a claim can be made when the NCO is confirmed. With non-agricultural land, a claim can only be made where notice to carry out a specified operation has been served and the person serving the notice has suffered losses or abortive expenditure as a result of delay caused by the statutory protection period.

Nature reserves

12.20 A nature reserve is defined by s 15 of the NPACA as land managed for the purpose:

(a) of providing, under suitable conditions and control, special opportunities for the study of, and research into, matters relating to the fauna and flora of Great Britain and the physical conditions in which they live, and for the study of geological and physiographical features of special interest in the area; or
(b) of preserving flora, fauna or geological or physiographical features of special interest in the area;

or for both purposes.

As with SSSIs, s 15 appears to proceed on the basis of current ecological value as opposed to potential ecological value but, again, English Nature appears to see potential value as a secondary criterion: see *Local Nature Reserves in England* (English Nature, 1991), p 8. Also, the definition appears to relate to the intrinsic value of the site; that is, the site itself must have the requisite ecological value and it is not enough to show that it has a value deriving from its contribution to other sites ('group value'), for example as part of a 'wildlife corridor'. Presumably the principles in *Sweet v Secretary of State for the Environment* (see **12.14**) apply, mutatis mutandis, and land of lesser value may be brought in as part of an overall site.

Designation

12.21 There are two types of nature reserve. A 'national nature reserve' (NNR) is designated by English Nature whereas a 'local nature reserve' (LNR) is designated by county councils, district councils, London boroughs and unitary authorities.

In terms of construing the phrase 'the area' in s 15, it is reasonable to argue that the 'area' for an NNR is the nation as a whole because English Nature will be acting on nationwide priorities. According to English Nature, 'the area' for an LNR is the administrative area of the relevant council; however, that leads to practical difficulties because ecological considerations do not observe administrative boundaries. It would be absurd if the subject site was on the edge of the area, or straddled the boundary, and land of ecological value only yards away in the next county or district were disregarded in assessing 'special interest'. Arguably, the administrative area of the acquiring council is a useful starting point, because the council will be approaching designation on the basis of priorities and a budget which covers the whole area. However, that starting point needs to be modified by reference to the particular facts of each case. In the *Royate Hill* case (see **12.24** below), the Inspector stated:

> '... it seems realistic to consider references to "the area" and "special" in this context as applying generally to the administrative area of the acquiring authority, whilst recognising that administrative boundaries have no ecological status and that whether something is special depends upon the context in which it is found. Hence it may be reasonable to attach greater significance to special opportunities within areas which are relatively deficient in them.'

Compulsory acquisition

12.22 Unlike an SSSI, designation of a nature reserve does not automatically create criminal penalties for any operations carried out on it. As with an SSSI, however, the objective is that the site will be regulated by a management agreement. If the owner is unwilling to conclude a management agreement and English Nature or the relevant council is worried about the future of the site, then the next step may be compulsory acquisition.

Under s 17 of the NPACA, English Nature, a county or district council, a London borough or a unitary authority may acquire the land compulsorily if they are satisfied that they are unable to conclude, on terms appearing to them to be reasonable, an appropriate agreement under s 16 provided that:

- in the case of English Nature, it is expedient in the national interest; and
- in the case of another relevant authority, that it is expedient in the interests of the locality (see s 21(4)).

In the *Royate Hill* decision (see **12.24** below), the Inspector distinguished the 'area' mentioned in s 15 from the 'locality' mentioned in s 21:

> 'However, I draw a distinction between "the area", which may be taken to be broadly equivalent to the administrative area of the acquiring authority and "the locality" (referred to in Section 21 of the Act). The latter term seems more equivalent to some concept of a neighbourhood, implying a more limited and less easily defined area within which people live, work and play.'

12.23 If the site is subject to a management agreement under s 16, then a breach of that agreement which prevents or impairs the satisfactory management of the land as a nature reserve will empower English Nature or the relevant council to issue a compulsory purchase order: s 18.

Derelict land

12.24 This power should be of major concern to potential developers of derelict land because a long period of disuse can lead to colonisation of a site by a variety of flora and fauna. The claim may then be made that the site ought to become a nature reserve.

In a letter dated 18 April 1995, the Secretary of State upheld a compulsory purchase order under s 17 in respect of a railway embankment at Royate Hill in Bristol that had been disused for over 20 years and had, in that time, become overgrown. During that time, local residents had trespassed on it and effectively used it as a local amenity. When the site was sold for development there was a local uproar which eventually led to compulsory acquisition of the site by Avon County Council. Nothing in the Inspector's report indicated that the site possessed any particularly rare species. He simply stated that it qualified as a nature reserve because it had 'some reasonable interest in the County context, and especially so in this urban locality': Department of the Environment ref SW/P/5041/12/2. Potential developers of derelict sites must, therefore, be extremely wary. If there is evidence that local people are using an overgrown site as an 'urban common', then there is potential for the use of this compulsory purchase power by the local authority.

Marine nature reserves

12.25 The WCA also makes reference to marine nature reserves; however, it is not intended to say much more about them in this book other than that they involve the NNR approach adapted to coastal and tidal areas.

PLANNING DESIGNATIONS

12.26 In the context of this book, SSSIs, NNRs and LNRs are of greatest significance when seeking planning permission. Such designations are a material consideration in the determination of planning applications and many development plans now have policies designed to restrain development which may adversely affect such sites of statutory conservation importance.

PPG 9 stresses that development proposals in or likely to affect SSSIs must be subject to special scrutiny (para 29) and Department of the Environment (DoE) Circular 15/88 (para 24) advises that consideration needs to be given to an environmental assessment where a Sch 2 project is likely to have significant effects on the special character of an SSSI. Paragraph 19 of PPG 9 states:

> 'Nature conservation objectives should be taken into account in all planning activities which affect rural and coastal land use, and in urban areas where there is wildlife of local importance. They should be reflected in regional planning guidance, structure plans, unitary development plans and local plans.'

Such development plans are required, as a matter of law (ss 12(3A), 31(3)(a) and 36(3)(a) of the TCPA) to include policies in respect of the conservation of the natural beauty and amenity of land and that requirement is reinforced by reg 37 of the Conservation (Natural Habitats & c) Regulations 1994.

12.27 In practice, development plans often create conservation policies which supplement statutory designations. For example, policy C4 of the Avon Structure Plan (Third Alteration) currently provides:

> 'Development which would affect adversely, either directly or indirectly, national and local nature reserves, sites of special scientific interest and other sites of nature conservation or geological importance normally will not be permitted.'

The structure plan does not define the phrase 'other site of nature conservation importance'. This category (then called 'other sites of wildlife importance') was provided by maps adopted by the council in 1987 (updated in 1991 and 1993 without public consultation) and by supplemental guidance entitled the *Avon Landscape Strategy*. This type of situation is worrying because such 'designation' can have dramatic and far-reaching effects.

Paragraph 14 of PPG 9 observes that 'our national wildlife heritage is not confined to the various statutorily designated sites' and para 15 goes on to note the importance of non-statutory sites which are given designations by local authorities and local conservation organisations. However, para 18 stresses that local authorities 'should only apply local designations to sites of substantive nature conservation value, and take care to avoid unnecessary constraints on development'.

The situation in Avon is now being rectified, in the sense that policy C4 sites are being defined in emerging local plans, but developers need to be very wary of their local situations, and should make representations upon newly emerging policies.

INTERNATIONAL OBLIGATIONS

12.28 It is necessary to briefly examine key international obligations in respect of nature conservation, because they have a practical impact on development projects in the UK; however, a detailed examination goes beyond the objectives of this chapter. Further advice on the topic can be found in PPG 9 and the DoE publication *Action for the Countryside*.

The Birds Directive

12.29 The UK is bound by the EC Directive on the Conservation of Wild Birds of April 1979 (Directive 79/409). The broad objective of the Directive is to provide for the protection, management and control of all naturally occurring wild birds in Europe, together with their eggs, nests and habitats. Member States are required to take measures to preserve a sufficient diversity of habitats in order to maintain the wild-bird population at an ecologically and scientifically sound level while taking account of recreational and economic requirements. The UK Government takes the view that those general requirements are met by current UK legislation and practices.

Article 4 of the Directive requires that, in the case of specified sensitive species, special protection areas (SPAs) be designated by Member States. If an area is to be classified as an SPA in the UK then, first, English Nature will designate it as an SSSI. English Nature will then recommend to the Secretary of State that it be further designated as an SPA. Owners and occupiers will be notified and will be allowed to make representations to the Secretary of State. The SPA designation does not, by itself, give greater day-to-day protection to an SSSI; however, it will be an additional factor to be taken into account when applications for planning permission affecting the SPA are determined and may have an effect on extant planning permission: see **12.33** below.

The Secretary of State is entitled to take into account economic factors in considering the designation of a site as an SPA: *R v Secretary of State for the Environment ex parte Royal Society for the Protection of Birds* [1994] JPEL B11.

The Bern Convention

12.30 The Bern Convention on the Conservation of European Wildlife and Natural Plants was ratified by the UK in 1983. This obliges signatory states to conserve wild plants and animals and their habitats, and lists species considered to be particularly vulnerable. Stress is laid on areas important to migratory species. The UK Government takes the view that the objects of the convention are achieved, in the UK, by existing designations such as SSSIs.

The Ramsar Convention

12.31 The UK is also a party to the Ramsar Convention on Wetlands of International Importance. This requires the government to conserve wetlands generally and to designate sites under the Convention. Again, the Government sees the SSSI system as being the starting point for Ramsar designations.

The Habitats Directive

12.32 Finally, there is the EC Directive on the Conservation of Natural Habitats and of Wild Fauna and Flora (Directive 92/43), which modifies the Birds Directive and provides protection to a wider range of habitats. The first broad objective of the Directive is to create a European network of habitats called 'Nature 2000'. Each Member State is to provide a list of potential special areas of conservation (SACs) with the objective of finalising the overall list by mid-1998. Each Member State must then designate sites on the final list of SACs. On 20 June 1995 Mr Gummer, the Secretary of State for the Environment, announced that he had, that day, sent a list of 136 candidate SACs to the European Commission. The list covered only sites on land and which were already SSSIs: see [1995] JPL 694–696.

The Conservation (Natural Habitats & c) Regulations are designed to transpose the Habitats Directive into national law. They provide for the designation of SACs and for the application of certain provisions of the Habitats Directive to SACs and SPAs.

International designations and planning control

12.33 If a site is designated as an SPA or SAC, then any proposed project having a significant effect upon it will require an environmental assessment before planning permission is granted: see para 39 of PPG 9. It is also likely that permitted development rights under the Town and Country Planning General Development Order 1988 will be restricted by an 'Article 4 direction'.

Regulations 50, 51, 55 and 56 of the Conservation (Natural Habitats & c) Regulations require LPAs to review extant planning permissions which are likely to have a significant effect on an SPA or SAC and, in appropriate cases, this may lead to the revocation or discontinuance of such planning permissions: see Annex C to PPG 9.

Chapter 13

ENVIRONMENTAL ASSESSMENT

INTRODUCTION

13.1 This chapter is designed to fulfil two purposes. First, to introduce the problems which arise in trying to establish whether the condition of a site is acceptable. Secondly, to advise on investigative techniques and obtaining site know-how. It will concentrate on the problems thrown-up by the redevelopment of 'brown-field' sites, but it will also form a useful introduction to the difficulties over standards which often arise when looking at specific regulatory regimes.

The broad statutory framework relevant to the environment assessment of certain planning applications has been discussed at Chapter 10; however, the matters discussed in this chapter will be directly relevant to the actual execution and content of such an environmental assessment. Surprisingly, there is a considerable paucity of accepted standards or guidelines for assessing when contamination exceeds acceptable levels. Therefore, in the majority of cases, specific proposals are judged on a pragmatic case-by-case basis and through negotiation with the concerned regulatory bodies.

THE SUITABLE FOR USE APPROACH

13.2 Before establishing standards, the body doing so needs to make a judgement as to the objective of the exercise. In both the USA and the Netherlands the objective is to require an owner to make the site fit for all foreseeable purposes. Thus it makes no difference if the developer wishes to use the site for a purpose which is relatively robust (eg a petrol station), he must make it fit for all purposes (eg residential, agricultural, etc). One has to question whether such an approach is realistic. Surely it would unnecessarily inhibit the beneficial use of otherwise derelict land?

The argument in support of this approach is that today's developer is making the land fit for future generations, but the criticism is that it also concentrates the burden of repairing past damage in one pocket. Aside from anything else, it contradicts the so-called 'polluter pays' concept because the present owner of the land is not necessarily the one who polluted it. Be that as it may, the approach in the UK is more pragmatic.

13.3 In the UK a proposal for the redevelopment of a contaminated site is currently tested by reference to the site's suitability for that proposed end-use – the 'suitable for use' approach.

The 'suitable for use' approach involves making a judgement as to the levels of pollution which can be accepted in relation to a specified end-use and then taking such action (if any is needed) which will bring the site within those levels. For example, a former industrial site might need no remedial work where the end-user is heavy industry, but it might need considerable remedial work if it is to be used for housing.

The 'suitable for use' approach has, however, been criticised by the House of Lords Committee on Civil Liability for Environmental Damage, which argued that it, 'condones continuing deterioration of our inheritance'. The report recommended that a phased programme of remediation be carried out over several decades and that 'the standards and priorities for restoration would be consistent with a resolve to pass to our children an inheritance free of danger and with as wide a range of possible land uses as can be achieved without excessive costs'. Unfortunately, there is only so much money available and it could be argued that the approach asserted by the House of Lords will lead developers away from contaminated urban sites and put pressure on green-field sites.

Planning Policy Guidance Note 23

13.4 Planning Policy Guidance Note (PPG) 23, entitled *Planning and Pollution Control*, was issued in July 1994. Paragraph 4.2 of the advice states:

> 'The Government's policy towards tackling the UK's burden of land contamination is that the works, if any, required to be undertaken for any contaminated site should deal with any unacceptable risks to health or the environment, taking into account its actual or intended use.'

Paragraph 4.3 of PPG 23 adds that the aims of the 'suitable for use' approach are, where practicable:

- to deal with actual or perceived threats to health, safety or the environment;
- to keep or bring back such contaminated land into beneficial use; and so,
- to minimise avoidable pressures on greenfield sites.

The Interdepartmental Committee on the Redevelopment of Contaminated Land (ICRCL) Guidance

13.5 The 'suitable for use' approach emerges in testing redevelopment proposals when obtaining planning permission and building regulation consent. ICRCL Guidance Note 59/83, *Guidance on the Assessment and Redevelopment of Contaminated Land* (2nd edn), states at para 6, that:

> 'contamination, or the potential for it, is a material planning consideration which needs to be taken into account at various stages in the planning process.'

That advice is now repeated, almost verbatim, in para 4.4 of PPG 23.

The Guidance Note deals with site investigations carried out prior to development and proceeds on the basis that it is laying down guidelines for providing a site which is suitable for its intended end-use. It does not apply to sites which have already been developed.

Table 3 of the Guidance Note provides 'tentative trigger concentrations for selected inorganic contaminants' (ie for the chemical contamination of soil). Table 4 includes guidelines for former gasworks sites. The tables provide 'threshold concentrations' and 'action concentrations'. These guideline levels are related to end-uses so that, for example, acceptable levels of contamination are lower for residential development than for commercial uses. The explanatory note to Table 3 explains its method thus:

> 'If all sample values are below the *threshold* concentrations then the site may be regarded as uncontaminated so far as the hazards from these contaminants are concerned and development may proceed. Above these concentrations, remedial action may be needed, especially if the contamination is still continuing. Above the *action* concentration, remedial action will be required or the form of development changed' (emphasis added).

The three guideline zones can be summarised thus:

- below threshold – no remedial action needed;
- between threshold and action levels – remedial action may be needed;
- above action level – remedial action needed.

Remarkably, however, Tables 3 and 4 do not give all of these very important action levels. In relation to 17 out of 34 examples, it is stated that: 'Action concentrations will be specified in the next edition of ICRCL 59/83'. Those action levels have not, as yet, been published and they include contaminants such as arsenic, lead and mercury. Therefore, ICRCL 59/83 does little more than provide some very basic guidelines. Indeed, arguably, it is potentially confusing in its current state because one could find regulatory bodies seeking remedial works or imposing constraints in all cases where the threshold concentrations are exceeded. Arguably, this has happened in at least one town and country planning case: see **11.10**. Furthermore, in 1990 the Environment Committee of the House of Commons criticised the ICRCL guidelines and there is a question as to their future.

The ICRCL guidelines are restricted to soil contamination and neither they nor any other UK standard deal directly with groundwater quality – albeit that there are standards for the quality of drinking water. Also, the guidelines are potentially confusing in relation to contamination from the subject site to neighbouring land or environmental media. According to the guidelines, the fact that a site is suitable for a given end-use does not mean that it has no potential to generate off-site pollution, for example, to contaminate groundwater.

Building regulations

13.6 Turning to the Building Regulations 1991 (SI 1991/2768), reg C2 provides that precautions should be taken to avoid danger to health caused by substances on or in the ground to be covered by the building. No measurable standards are provided, therefore the application of reg C2 is a matter of opinion in each case. It also needs to be observed that there are some types of development which do not need building regulation approval in any event.

It is also worth noting that the National Housing Building Council (NHBC) may, if a site is 'hazardous', require a full site survey and appropriate design modifications before granting their registration certificates. Their interpretation of 'hazardous' is defined in Appendix 4.1A of the NHBC Standards.

FOREIGN STANDARDS

13.7 One has to ask whether, given the paucity of UK standards, it is sensible to look at standards applied in other countries.

The Netherlands standards

13.8 In Europe, apart from the UK, only the Netherlands has adopted any standards – and they cover both contaminated soil and water. The Netherlands standards make no reference to background conditions or site-specific conditions and are designed to suit the geological conditions in Holland, where gravels and sands are common with a high water table causing rapid movement of contamination in groundwater. By contrast, many areas in the UK are underlain by relatively impermeable clays which (in the absence of underground fissures or penetrations) slow or hold the movement of contamination. In the Netherlands, about 80% of national drinking water comes from groundwater supplies – compared to a corresponding 35% of public water supply in England and Wales.

Canadian standards

13.9 The Netherlands standards formed the basis for the *Interim Canadian Environmental Quality Criteria for Contaminated Sites* produced by the Canadian Council of Ministers for the Environment (CCME) in 1991 – and, in turn, these federal criteria form the basis for standards within the provinces of Canada.

The CCME adapted the Netherlands standards to suit Canada's own unique geology and are currently developing a database which will allow for regional variations. One of the Canadian approaches to contamination (ie that in Alberta) involves using provincial standards based on the CCME criteria as a starting point, but then going on to analyse site-specific conditions and local geology to make a risk assessment. The question of whether or not the project goes ahead, and in what form, is thus decided by reference both to objective guidelines and expert judgement.

THE RELATIVE NATURE OF CONTAMINATION

13.10 The assessment of acceptable levels of contamination will inevitably always turn on case-by-case judgments because sites can vary dramatically in characteristics and location. A local planning authority might be happy to permit the construction of a petrol filling station in an industrial area, but less happy to do so if it is to be located over an important drinking water aquifer; and proposals to redevelop a methane-producing closed land-fill site might be more readily received if the site is away from a residential area – regardless of assurances about gas control.

Different considerations will apply where a site is based on impermeable material as opposed to highly permeable material, because relative permeability affects the relative movement of contaminants. There will also be differences between sites due to the depth at which the contamination is located and differing soil chemistry.

Given these considerations, it is not surprising that there is considerable variation between various regulatory bodies around the country. The second report of the House of Commons Environment Committee on toxic waste criticised this aspect of the operation of the waste-management provisions of the Control of Pollution Act 1974, characterising the inconsistency of standards between waste-disposal authorities as a *laissez faire* approach. Perhaps there was a case for criticism, but the reality is that regulatory bodies have to respond to local geography, geology and population patterns.

OBTAINING KNOW-HOW

13.11 Regardless of the difficulties one may find in identifying suitable standards or guidelines, the fact remains that those involved in land have to take into account issues relating to known or possible contamination. Therefore, whenever any person is likely to become involved in a land transaction, he ought to carry out prudent checks as to the environmental condition of the land itself and, perhaps, neighbouring land. That is not to say that a comprehensive and highly expensive site investigation should be carried out in every case, because there are a number of levels at which an environmental assessment can be carried out.

There is no established UK national standard for investigating the condition of sites. The closest is draft standard DD175: 1988 produced by the British Standards Institute (BSI) and entitled *Draft for Development: Code of Practice for the Identification of Potentially Contaminated Land and its Investigation*. The draft Code was put forward as provisional guidance only on the basis that it would be reviewed in 1990, but that review has not, as yet, taken place.

13.12 The draft Code of Practice provides useful guidance in relation to desk studies and intrusive site investigations (see below for the meaning of those expressions). It is designed for the identification of potentially contaminated sites and assessing the suitability of such sites for the intended use, and relies primarily on the 'desk study' technique and intrusive site investigations.

The Department of the Environment (DoE) has assisted in environmental assessments by the publication, at the end of 1994, of a series of Contaminated Land Research Reports. Each report has been prepared by a consultant with particular expertise in his topic area – these reports can be purchased from the DoE (Publication Sales Unit, Block 3, Spur 7, Government Buildings, Lime Grove, Ruislip, HA4 8SF) and represent good value for money.

Some guidance can also be gained from the US procedures for environmental site assessments which underpinned the 'innocent purchaser' defence to the Comprehensive Environmental Response Compensation and Liability Act 1980 (ie 'Superfund') liability. The defence can be raised where the purchaser proves that he has carried out 'all appropriate inquiry into previous ownership and uses of the property consistent with good commercial or customary practice'. There has been considerable debate as to how this defence can be achieved and the American Society for Testing and Materials (ASTM) produced, in May 1993, two standard procedures for the environmental assessment of commercial (ie non-residential) property; namely, the 'transaction screen' and the 'phase 1 environmental site assessment'. Both processes are non-intrusive in the sense that they do not involve any bore-holes, trial-pits or other engineering operations or any testing or sampling of materials. The draft BSI Code of Practice does not contain anything which parallels transaction screening, but does go further than the ASTM in one area as it gives detailed guidance on intrusive site investigations.

Intrusive site investigations will be discussed below, together with the relatively economical 'desk study' technique. The ASTM 'transaction screen' procedure will not be discussed because it is not widely used in either the USA or the UK.

The desk study

13.13 A 'desk study' is, primarily, documentary research in relation to the subject site and its locality which is best carried out by an environmental professional. Both the ASTM standard and the draft BSI Code of Practice contain guidance on desk studies. The most comprehensive guidance is the DoE Contaminated Land Report, *Documentary Research on Industrial Sites* (CLR No 3). Whilst the phrase 'desk study' is used in this book, this phase of assessment may also be described as a 'record review' or 'documentary research'. The investigator will collect data from sources such as historic ordnance survey maps, aerial photographs, planning records, trade directories, rating records etc, and make a recommendation deriving from that exercise. CLR No 3 (para 2.1) describes the purpose of the study as being research on the history of a site's industrial and other past uses in order to:

- establish the likelihood of contamination being present and whether further site investigation is required;
- establish whether any special precautions are needed during site inspection or investigation; and
- provide information required in the design of site investigation and/or remedial measures.

A fourth purpose needs to be added to that list; namely, to provide information to be used in the sale and purchase of the site. All substantial property transactions ought to involve a desk study as a matter of routine.

The information most commonly sought through documentary research to meet these objectives is listed by para 2.2 of CLR No 3 as being:

- the history of industrial and other uses of a site;
- the processes used including raw materials, products, waste residues and methods of disposal;
- the layout at each stage of a site's history, including locations of materials handling areas, buildings, underground structures, services, etc; and
- evidence suggesting the presence of wastes or other materials.

The investigator should explain the range and depth of his record review because records vary in their ease of examination and he will not undertake to review every possible record which may exist in relation to the property. Unless the study has been commissioned on a 'property only' basis, the investigator ought to review the physical setting of the property (that is its geology, hydrology, hydrogeology and topography) by reference to readily available data and explain how he arrived at his geographical area of search. The investigator will normally review records from standard sources only and, under the ASTM standard, only information which is reasonably ascertainable from these sources. The information is 'reasonably ascertainable' if it is available to the public, obtainable within reasonable time and cost restraints and is in a form which yields information relevant to the property without the need for extraordinary analysis of irrelevant data. The BSI draft Code does not discuss this issue.

CLR No 3 provides extremely full advice both on documentary sources and the way in which they can be interpreted, in particular as to the use of ordnance survey maps. It contains a detailed list of supplementary map sources including Goad fire insurance plans, Parliamentary enclosure maps, plans of public and statutory undertakings (eg drawn up for the construction of canals, railways or public utilities), tithe maps, town maps and plans, British Geological Survey maps and Soil Survey and Land Research Centre maps. In terms of collated documentary sources, advice is given in relation to trade directories (eg Kelly's), statutory planning registers, records held by waste-disposal authorities, aerial and satellite photographs, the DoE surveys of landfill sites (1973/74 and 1988) and the partial survey of contaminated land in Wales carried out by the Welsh Office in 1988. Further assistance on this aspect can be obtained from Chapter 15 of this book.

The preliminary environmental assessment

13.14 There is no universal understanding of the phrase 'preliminary environmental assessment' or allied phrases such as 'phase 1 environmental assessment'. The BSI draft Code of Practice sets a standard for what is described as a 'preliminary investigation' and the ASTM sets a similar standard described as the 'phase 1 environmental assessment'. Both approaches have the following four components:

- a desk study;
- a reconnaissance of the site and its locality;
- interviews with current owners and occupiers and government officials; and
- an evaluation and report.

The assessment will be carried out by an environmental professional. A desk study or records review follows the desk-study techniques described above. The DoE Contaminated Land Research Report, *Guidance on Preliminary Site Inspection of Contaminated Land* (2 vols) (CLR No 2) gives extensive guidance on the conduct of the reconnaissance – there called the 'preliminary site inspection'.

The reconnaissance involves systematic observation of the site, adjoining uses and the surrounding area. The site reconnaissance will be on foot and in detail whereas the reconnaissance of the locality may be from a car or public rights of way and will be more broadly based. The investigator will look at various indicators of potential contamination. Gaps in vegetation may indicate the presence of toxic substances. One would normally expect to see vegetation on an unused site and its absence, or erratic distribution, indicates the need for further investigation: see paras 3.10.1 and 3.10.2 of CLR No 2 (vol 1). Unusual colours in surface materials may be due to the presence of chemicals, for example yellow colouring may indicate the presence of chromium, and black or orange/brown may indicate severe landfill-gas contamination: see para 3.7 of CLR No 2 (vol 1). It may be possible to detect unusual smells and, in some cases, there might be debris on the site indicative of past uses, for example old drums. The landform of the site may be an indicator. Contours may appear to be artificial or discontinuous with the surrounding area thereby suggesting the possibility of tipping, quarrying and so on. Under both the ASTM standard and the BSI draft code, the investigator is expected to make a reasonable attempt to interview the owner and occupiers of the property (either in writing, in person or by telephone) about the past and current uses and environmental condition of the property and to pose similar questions to appropriate government officials. Whilst anecdotal evidence has to be treated with caution, personal recollections of those involved with or neighbouring the site can help explain a complicated site history or indicate unrecorded past working practices.

13.15 Enquiries to owners and occupiers include obtaining information on plant construction and layout, process operations, waste disposal, and asking to see business records, leases and deeds. Enquiries of government officials can include local authorities, waste disposal authorities, statutory undertakers, Her Majesty's Inspectorate of Pollution and the Factory Inspectorate. The type of information sought can include details of the planning history of the site, its environmental health and building control records, information on the construction, layout and use of drains and sewers, and gas, water and electricity mains. Details of past industrial operations can also be sought: see generally Table 2 of DD175: 1988.

The evaluation and report will identify potential hazards on or to the site and those things which are or would be potentially at risk. It should also advise on the safe and proper conduct of further investigations and on any remedial measures which need to be taken immediately or before further investigations. The Welsh Development Agency (WDA) *Manual on the Remediation of Contaminated Land* sets out a useful summary of the content of a typical preliminary investigation report. This is reproduced at Appendix 2(3) of this book.

The main environmental assessment

13.16 The main environmental assessment may either be 'once and for all' or, more likely, it may take place in stages to suit either the phased nature of the development or the investigator's need to gather and evaluate data in successive cycles. It will probably be based on intrusive site investigations, but may also include gas and vapour surveys and geophysical techniques. Its main parameters ought to have been established by a preliminary environmental assessment, but its first stage might involve further preparatory investigation and testing to refine existing preliminary information.

13.17 No site assessment will be completely exhaustive in the sense of eliminating all uncertainty about site conditions. A balance has to be struck between the costs of the investigation and the value of the information being obtained. One can quickly get into a situation where the usefulness of the information is outweighed by the cost of obtaining it and this has to be borne in mind when structuring the investigation. It must also be stressed that there is no standard method of carrying out a main site assessment. The BSI draft Code of Practice deals with intrusive site investigations, but the ASTM standard does not.

An intrusive site investigation involves obtaining samples of soil, sub-soil, gases and groundwater, by bore-holes or trial-pits, with reference to a predetermined grid-pattern over the site and testing them. The samples will include both contaminants and the environmental media on the site. If the site has buildings or hard-cover on it (eg old foundations), then it may be necessary to engage in demolition or site clearance before the investigation can take place. If the site contains a body of water (eg a pond, lake, lagoon etc), then it may be necessary to take samples from a boat or raft. The investigator may also take cuttings from on-site plants for analysis and, in some cases, collect and test fauna.

If the investigator does not have information about the history of the site or the proposed end-use, then he will have to work according to a sampling system designed to predict the condition of the site as a whole. Since the investigator would not, in this situation, have any advance indication of the contaminants which would be harmful to the final use of the site, he would test for a wide range of contaminants, many of which would, in the event, prove to be irrelevant.

13.18 The various approaches are discussed in the DoE Contaminated Land Report, *Sampling Strategies for Contaminated Land* (CLR No 4), and illustrated in Figures 1 and 2 below. According to CLR No 4, an efficient sampling pattern should satisfy four conditions:

(a) it should be partitioned into regular sub-areas;
(b) each sub-area should carry only one sampling point;
(c) it should be systematic; and
(d) sampling points should not be aligned.

The CLR advises that the *simple random* pattern satisfies condition (d) only; the *square grid* pattern satisfies all but condition (d); the *stratified random pattern* satisfies all but condition (d); but the *herring bone* pattern satisfies all four conditions and has a better performance in detecting elongated 'hot-spots' (see below).

13.19 In practice, one would not expect to see the use of a sampling pattern across a site in a completely non-discriminatory way. Aside from the question of unnecessary expense, there are two practical reasons.

First, the levels and types of contamination will probably vary across the site, thus requiring that some areas ('hot-spots') be sampled intensively whereas other areas may be relatively innocuous. This type of unequal sampling can be undertaken when a preliminary investigation has given strong guidance as to the probable distribution and nature of contamination on the site or has revealed possible problem areas (eg old underground storage tanks).

Secondly, unless the proposed development is to be completely homogeneous across the site, different elements of the development will have differing sensitivities to, and effects on, contamination. A hard surface, such as a car-park, is unlikely to be sensitive to most forms of contamination but, on the other hand, the contamination may be sensitive to it – in that the contaminant may produce gases that would otherwise emit to the atmosphere, or the hard surface may adversely affect the migration pattern of underground contaminants. If the contaminant emits gases (eg a landfill site), then the construction of buildings which can trap the gases may be a problem. Or the construction of a building in a particular spot may place it over a contaminant which is aggressive to its foundations or which may

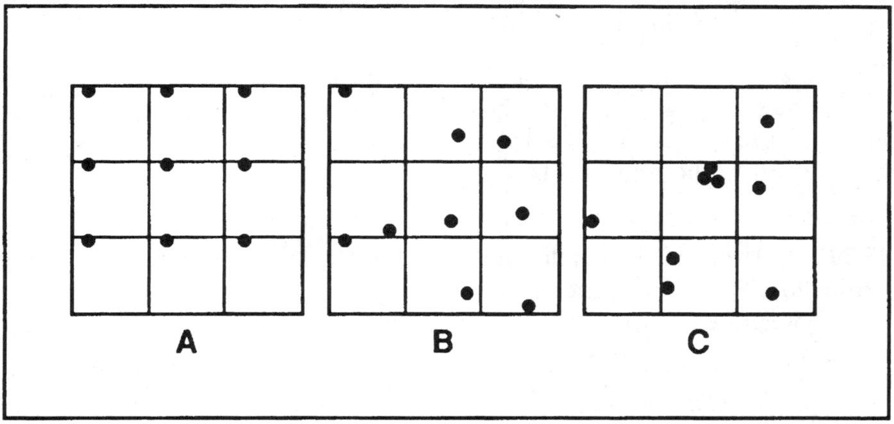

Figure 1 Three sampling designs: A, regular square grid; B, stratified random; C, simple random*

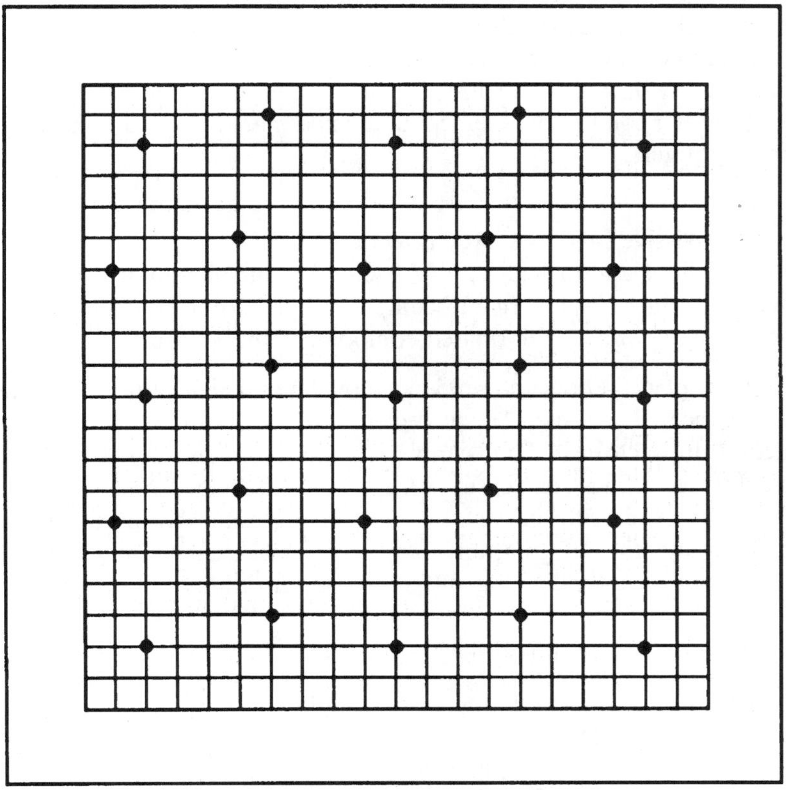

Figure 2 Herringbone sampling design*

*Source: CLR No 4 (DoE, 1994)

damage mains services to it. Conversely, an end-use such as open landscaping might not be sensitive to or adversely affect contamination.

Thus, there is no single universal formula which can be applied from site to site. If the site investigation is to be carried out in an effective and economical manner, then it must be preceded by a preliminary site assessment and must have regard to the proposed layout of the development.

13.20 A large-scale or intensive site investigation may need planning permission. Some trial trenching can be 6 m deep and require the services of a heavyweight excavator – this would probably be 'engineering' development requiring planning permission: see s 55 of the Town and Country Planning Act 1990. In *Bedfordshire County Council v CEGB* [1985] JPL 48, the Court of Appeal held that the drilling of three boreholes (60 ft deep × 2 in diameter) on a 400-acre site (to see if it was suitable for disposal of nuclear waste) did not require planning permission, but the court seemed to assume that a more intensive investigation would. It may also be necessary to comply with a multiplicity of safety legislation including the Health and Safety at Work etc Act 1974. It is also worth ensuring that the contractor carries proper insurance. Sometimes even the most carefully carried out investigations may cause damage – for instance, if a borehole strikes an underground pipe or creates a conduit between contaminated material and an underground stream or aquifer.

13.21 The passive soil gas survey is a cost-effective complement to intrusive site investigations and provides a rapid screening of soil and groundwater for volatile and semi-volatile organic compounds including landfill gas, petroleum-based compounds, chlorinated compounds and solvents.

Samplers are placed about the site in shallow holes which cause the minimum of disturbance. They can be placed in buildings and will not disrupt production in factories. The samplers remain on site for several days, continually collecting soil gases which are diffusing from underground sources. At the end of the survey, the samplers are returned to the laboratory for analysis and the information obtained is then used to produce colour-coded contour maps indicating different concentrations of the gases being investigated.

13.22 Geophysical investigations can involve any of a wide range of techniques from magnetic and gravimetric measuring to infra-red photography. Again, these are complementary to intrusive investigations and will help to quickly identify the relevant parameters of the main investigation. They are based on the fact that land which has been contaminated has differing physical properties from nearby land which is in its natural state. Thus, the contaminated area might have different heat-retention properties to its surroundings or decomposing organic waste may be warmer, and these phenomena may be detectable by infra-red photography.

The WDA Manual helpfully provides examples of common investigation techniques and they are reproduced at Appendix 2(3) of this book. Appendix 2(3) also reproduces the Manual's summary of the minimum content of a final investigation and assessment report.

13.23 It is not possible to escape the fact that a comprehensive site investigation could be very expensive. As such, it is very important that these costs are incurred at the most appropriate time in the transaction. For example, if the transaction involves obtaining planning permission, the prudent developer would try to defer a detailed site investigation until after outline planning permission has been obtained – thereby establishing that the development is accepted in principle before incurring major investigative costs.

One often finds that a prospective purchaser or developer will require site investigations for ground engineering, geotechnical and archaeological purposes, therefore it will often be economical to combine these investigations with contamination investigations. The main site assessment must investigate, and take into account, the physical character of the site and its locality and the potential behaviour of any contaminants in relevant environmental media (eg groundwater) or through potential pathways (eg old mine workings or wells).

GROUNDWATER AND SURFACE WATERS

13.24 The significance of groundwater contamination has recently been thrown into prominence by the *Cambridge Water Co* case (see **5.22**). DoE Contaminated Land Research Report, *A Framework for Assessing the Impact of Contaminated Land on Groundwater and Surface Water* (2 vols) (CLR No 1) provides detailed advice on the assessment of the impact of contaminated land on groundwater and surface water, and one would expect that advice to be reflected in the main environmental assessment. In very broad terms, CLR No 1 advocates a 'source – pathway – receptor' analysis. This is illustrated below. As its title suggests, the approach begins with an analysis of the site (ie 'source') in terms of its potential for groundwater contamination. Factors to be taken into account include the level and type of contamination and the way in which precipitation (ie rainfall and snow) affects the site, for example through evaporation, absorption by the soil, infiltration into deeper deposits and trace groundwaters, and run-off into surface waters.

13.25 Analysis of the 'pathway' involves examination of the surface of the site and its underlying geology. If the site has large areas of hard surfacing or the moisture absorption characteristics of the soil are low, then migration of contaminants in run-off is a possibility. If the site is situated above a porous granular medium, such as sand, then the migration of contaminants is a possibility. If the pathway is by way of natural or man-made conduits (eg fractures, fissures, old mineshafts and wells) then migration may be comparatively rapid. Analysis of the receptor involves assessing its ability to cope with pollution. A fast-running river may have a greater capacity to

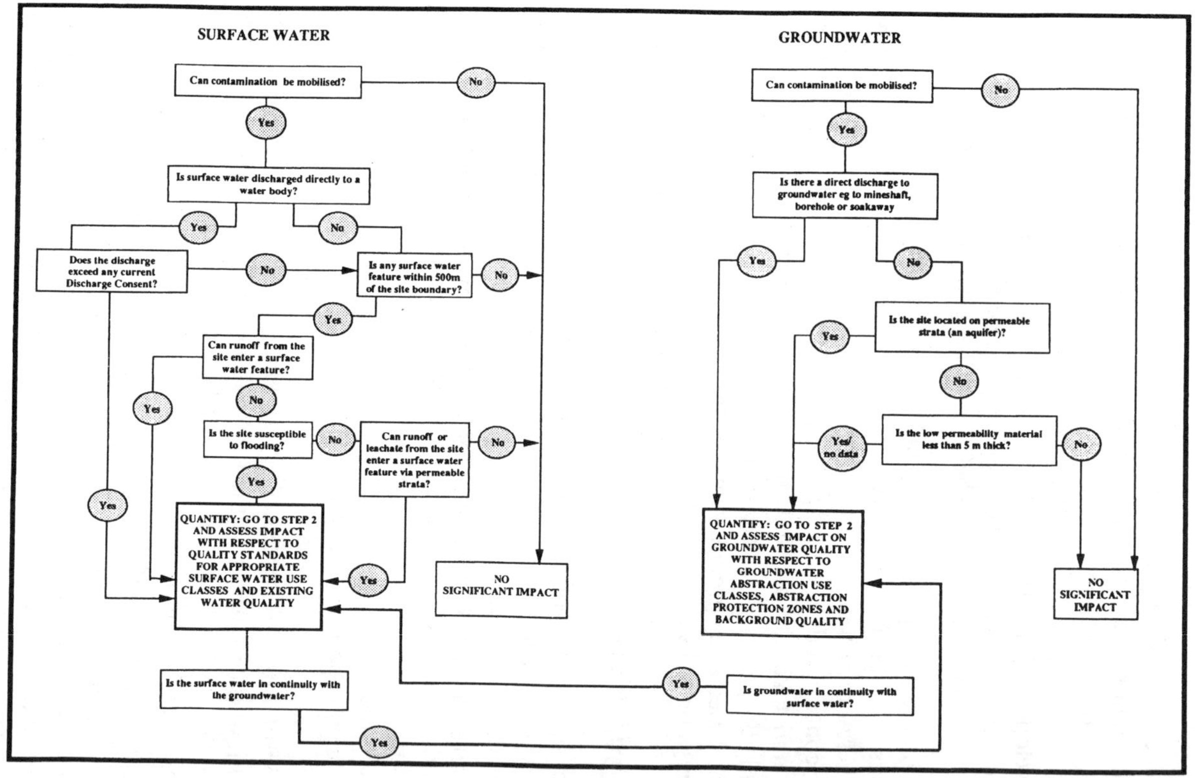

Figure 3 Assessing the impact of contaminated land on groundwater and surface water: step 1*

*Source: CLR No 1 (DoE, 1994), vol 1

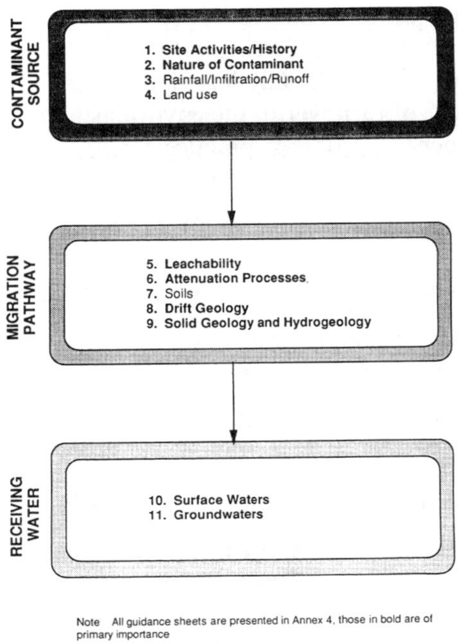

Figure 4 Assessing the impact of contaminated land on groundwater and surface water: step 2 – information needs and guidance sheet numbers*

*Source: CLR No 1 (DoE, 1994), vol 1

disperse a given level of contaminant than a slow-moving groundwater source or a static pond. The broad principles involved can be seen at Figures 3 and 4 above.

CHOOSING THE EXPERT

13.26 There is no widely recognised standard or certification system for those involved in site assessment; therefore, considerable care needs to be exercised in choosing a consultant. A choice needs to be made having regard to expertise, experience and financial ability to carry the risk.

In October 1993, the Government awarded the National Accreditation Council for Certification Bodies the task of accrediting organisations which will certify compliance with BS 7750, the EC Eco-Management and Audit system and the environmental management systems standard: Hansard, HC Written Answers col 198, 28th October 1993. This will go some way to the provision of officially recognised environmental auditors. One of the organisations involved in this emerging standardisation of qualifications is the Environmental Auditors Registration Association (EARA), which was launched in 1992. Registration with the EARA depends on academic and professional qualifications, training and practical experience. By late January 1994, over 500 auditors were registered with the EARA.

13.27 In order to bring some certainty to the process of appointing consultants, the Association of Environmental Consultancies (AEC) was set up in 1991. Currently, the AEC only has 28 member firms; however, they are said to represent 60% of turnover in the environmental consultancy sector. In 1994 the AEC launched a code of practice for its members and clients will have access to data on firms registered under the scheme, including details of staff qualifications and experience, indemnity cover and past clients: see ENDS Report 234, p 14.

The consultant should be able to offer expertise in the following disciplines:

- information research and retrieval;
- environmental science;
- environmental legislation and regulations;
- chemistry and geochemistry;
- geology and hydrogeology; and
- engineering (civil/chemical/mechanical).

Site assessment has a relatively short history, therefore most firms will only be able to offer 5- or 6-year track-records. However, an active firm will have dealt with hundreds of projects during that period and should be able to supply details together with appropriate corporate references. The sensible client would not take the consultant's word for past performance without appropriate checks.

It is necessary to check whether the proposed consultant can carry the risk of any error or omission. Unless the firm has demonstrably massive financial resources itself (and it is proved that they will remain so for decades), the only reasonably reliable method is by way of insurance. Most ordinary professional indemnity policies now exclude environmental liability, and policies which do not will carry heavy premiums. Only a consultant with full coverage should be selected. A recent survey has indicated that almost 60% of the environmental consultancies with professional indemnity cover which responded were not insured against claims arising from gradual pollution incidents: ENDS Report 241, p 3. Also, client confidentiality permitting, it is worth seeing the type of report typically produced by the proposed consultant.

13.28 Finally, it is worth noting that the actual analysis of samples is far from an exact science. In 1993 the Laboratory of the Government Chemist (LGC) launched a scheme called 'CONTEST' to provide laboratories accredited for the analysis of soil contaminants. The LGC noted at that time that results from two good quality laboratories could vary by 20–30%.

There is, as yet, no British Standard regime for soil contaminant testing, and as such it is important to be very discriminating in selecting an appropriate laboratory.

Chapter 14

THE MANAGEMENT AND DISTRIBUTION OF RISKS

INTRODUCTION

14.1 Transactions in land always involve some form of risk-taking and risk-distribution. By its nature, contaminated land will always be something of an unknown quantity. There is no guarantee that even the most thorough site investigation will fully predict a site's potential for pollution. Furthermore, there is the perennial problem that there is a paucity of clear-cut standards in the UK by which to measure a site's level of contamination. Even then, today's standards are not tomorrow's standards – and new standards are usually tighter.

14.2 Inevitably, the parties to a transaction must decide who is going to bear particular risks. That might be by way of the direct acceptance of a risk or by way of an indirect responsibility through an indemnity, warranty or the like. Another approach might be to bring in a third party (eg an insurer, guarantor or bondsman) to accept certain risks.

This chapter will begin by identifying the particular risks which may be faced by lenders, freehold purchasers, leaseholders and receivers. It will then explore some of the strategies and devices which are generally used to manage and distribute risks.

CONTAMINATED LAND AND THE LENDER

14.3 The position of lenders under current legislation has always been ambiguous. There has always been concern that the regulatory bodies would try to implicate them in the misdeeds of their often impecunious borrowers in order to ensure that at least one party to an enforcement action would be in a position to meet the bill. Recent statements from the Department of the Environment (DoE) have not given lenders any reasons to feel confident about the future.

On 13 October 1993 the then Environmental Protection Minister, Mr Tim Yeo, made a speech to the British Bankers' Association which caused considerable alarm. He noted that 'lenders have the power to influence industry's behaviour' and went on to say:

> 'In the complex modern world, it is not enough to say the polluter must pay – and make that an excuse to leave it to the last person in the chain of decisions. Liability should follow responsibility – possibly even to bankers under some circumstances. Otherwise, there is every incentive to make the

> operator of a polluting process into a "man of straw" who can leave the public purse to bear the costs of cleaning up his mess.'

The DoE consultation paper entitled *Paying for Our Past* (March 1994) did little to dispel that alarm. Whilst purporting to follow the 'polluter pays' principle, it indicated, as a preliminary conclusion, that:

> 'People or organisations should normally be held liable for any responsibility under statute which they might have, even if indirect, for contamination. Accordingly, regulators should – in so far as appropriate – be able to pursue *lenders*, financial institutions, etc. for recovery of costs in cleaning up contamination' (p 26) (emphasis added).

This statement has been amplified, with regard to lenders, by the Government's policy document *Framework for Contaminated Land* which was published in November 1994. Paragraph 4.4.10 states:

> 'Lenders should not expect to be relieved of all risk in relation to the sums they advance. Nor should they, or other persons exercising control of land or a business, be relieved of responsibility where, in exercising their functions, they had caused or contributed to the damage. They should also take proper care in entering into the roles specified and in the way they carry them out, including their material influence over the actions of clients, associates or agents.'

And para 4.4.12 adds:

> 'But regulatory authorities should not be able to treat financial institutions as "deep pockets", being made liable for the costs of any remedial works regardless of any responsibility.'

The 'Framework' concludes that the existing regulatory laws concerning persons who 'cause or knowingly permit' damage to other land or to groundwater provide a proper balance in respect of lenders.

Lender liability: the control test

14.4 It is instructive to look at the way in which the USA has dealt with the problem of lender liability in the context of its Comprehensive Environmental Response Compensation and Liability Act 1980 (commonly known as 'Superfund' because it set up the 'Hazardous Substance Superfund' to pay for remedial works in respect of pollution).

The 1980 Act places strict retroactive liability for, among other things, damage arising from the release of hazardous substances on the 'owner or operator' of the contaminated site but excludes a person who 'without participating in the management of [the] facility, holds indicia of ownership primarily to protect his security interest in the vessel or facility'. Many lenders in the USA thought that this exclusion meant they would be immune from remediation costs, under Superfund, when they entered into possession. However, *United States v Fleet Factors* (1990) 724 F Supp 957 caused them to think again. There the US Supreme Court held that a lender was liable for remediation costs if he participated in the day-to-day functionings of a company, that is where he had the capacity to influence the company's treatment of hazardous waste. The enforcing body, the US Environmental Protection Agency, provided an interpretation of the lender exception in April 1992 to allay some of the fears generated by *Fleet Factors*, but the statutory basis for the exemption remains unchanged: see Groy & Cahill (1993) *Env. Liability*, vol 1, p 1.

14.5 The US approach gives some guidance in the context of UK legislation on liability for remedial works. If a mortgagee is *out of possession* of the land then, in effect, the mortgage is merely by way of a security. As such, one could argue that despite a mortgagee's technical status as having an interest in land, it is not the type of interest contemplated by statutory remedial powers. A mortgagee *in possession* is, almost certainly, an occupier and, on a control test, may be an 'owner'. He could therefore find that not only has he the liability arising from his unpaid loan, but also the liability for remediation costs.

14.6 If a receiver acts in accordance with the instructions of a debenture holder or mortgagee then he may be deemed to be the agent of the lender instead of the agent of the borrower: see *Standard Chartered Bank v Walker* [1982] 3 All ER 938; also *American Express v Hurley* [1985] 3 All ER 564. Thus a lender who fails to keep at arm's length from a receiver could become liable for the receiver's conduct in respect of charged property.

The lender's options

14.7 One approach is to exercise considerable control over what the borrower may or may not do with the site. If the borrower's operations are inherently contaminative, then the lender could require that environmental audits be carried out (and reported to the lender) or even that the borrower participates in an environmental management regime such as BS 7750 or the EC's Eco-Management and Audit Scheme.

14.8 Lenders should also ask themselves, in relation to environmental liability generally, whether their loan documents should include a power to enter the site to carry out pre-emptive works designed to prevent the build-up of future environmental problems. The difficulty with this approach is that the lender might then be treated as having sufficient possession to be an operator, occupier or owner – with all that entails. Another possibility is that the lender is given power to close the site or re-evaluate the terms of the loan where contamination or liabilities arising from the site exceeds certain thresholds.

A lender should always have access to any relevant environmental information held by the borrower including licences, permits and planning permissions. An imaginative adviser can create a plethora of devices designed to protect the lender, but one needs to be aware that they arrogate, step by step, control to the lender.

THE FREEHOLD PURCHASER

The caveat emptor rule

14.9 The purchaser of a freehold has, first, to take such steps as are prudent to try to ensure that either he is not buying a contaminated site or the level of contamination is made known to him before he becomes irrevocably committed to the purchase. Site-assessment procedures are discussed in Chapter 13, but what duties of disclosure lie on the vendor in the first instance in relation to the quality of land? The seller of land in England does not, normally, have to make any disclosure to the purchaser about any defect in the land. This is known as the 'caveat emptor' principle.

14.10 The Conveyancing Standing Committee of the Law Commission considered the caveat emptor principle in 1988. They listed 12 exceptions to the principle including:

(a) Contracts of utmost good faith.
(b) Where a fiduciary relationship exists between the parties (eg solicitor and client).
(c) Where a positive statement by the vendor would be distorted by silence, as where essential facts are omitted. For example, in *Lawrence Ltd v Lexcourt Holdings Ltd* [1978] 1 WLR 1128, where a planning permission was incomplete, and in *Shepherd v Croft* [1911] 1 Ch 521, where the vendor said that 'several excellent building sites could be formed', but neglected to mention that an underground watercourse would be a source of major structural problems: see also *Sidney v Buddery* (1949) 1 P&CR 34.
(d) Where the vendor has caused the problem himself, as with the defective 'do-it-yourself' work in *Hone v Benson* (1978) 248 EG 1013. One can easily see the application of this exception in the context of contaminated land, for example where the purchase is from the original contaminator.
(e) Where the vendor has concealed physical defects either by false representation or by works of concealment, provided that the purchaser relies on that concealment: *Gordon v Selico Ltd* [1986] 1 EGLR 71.
(f) The Committee also opined that there may be a duty on the vendor to reveal a physical defect known to him which would endanger life. This could be an important principle in the current context but the Committee accepted that there is little authority for the proposition.

The Conveyancing Standing Committee described caveat emptor as 'an unjustified ramshackle principle' which was 'suffering from the legal equivalent of subsidence'. Therefore, they recommended that the principle should be replaced by a positive duty on the seller to disclose all material facts about the property he is selling; however, they backtracked in 1989 with a somewhat vague recommendation that voluntary arrangements under which standard basic information is volunteered to buyers should be encouraged. The House of Commons Environment Committee looked at the principle in 1989/90 and failed to make any positive recommendation. The current position, and the position for the foreseeable future, is that broadly, 'let the buyer beware'.

14.11 It is, also, worth bearing in mind the Property Misdescriptions Act 1991 which makes it an offence '... to make a false or misleading statement about a prescribed matter in the course of an estate agency or a property development business'. 'Prescribed matters' include the physical characteristics or condition of the land or its fitness for purpose; treatments or processes to or on it; and surveys, inspections, investigations or appraisals. Absolute silence is not caught by the Act, but vendors of property are rarely completely silent when trying to sell it. The Act does not create civil liability for misdescriptions.

14.12 However, there is rarely any excuse for a purchaser's failure to take elementary steps to obtain knowledge of the condition of the relevant site. Clearly, the costs of taking those steps need to be commensurate with the value of the transaction. There is no reason why, at the very least, all freehold purchases of substantial value should not be preceded by a desk-study: see **13.13**. All conveyancing transactions should involve environmental preliminary inquiries before contracts are exchanged: see Model 1, Appendix 1.

Undertakings comprised in indemnities, warranties and insurance are discussed later in this chapter. Whilst these undertakings have clear limitations, they will normally form part of every substantial transaction because it will never be possible to obtain complete and perfect information as to the condition of the site.

Freehold purchasers: options and conditional contracts

14.13 Many commercial acquisitions now proceed by way of options to purchase and conditional contracts. An option is a contract between vendor and purchaser whereby the purchaser is entitled to call for the transfer of the freehold upon the happening of a defined event – often the grant of satisfactory planning permission. A conditional contract requires the purchaser to complete purchase of the property upon the happening of a specified event.

14.14 The implications of contaminated land can often be overlooked in this type of transaction. For example, the purchaser under an option may covenant to use his best endeavours to obtain a satisfactory planning permission and then find that, due to the likelihood of contamination, the local planning authority (LPA) requires an expensive site investigation to be mounted. The cost of the site investigation may outweigh the value to the purchaser of the transaction, yet he is bound to carry it out by the terms of the option. Arguably, the purchaser should have accepted a 'reasonable endeavours' commitment only, but the better solution is to provide that he will not be liable to carry out onerous surveys.

To give another example, a conditional contract may be complete on the grant of a specified planning permission, yet that permission may be subject to a condition requiring that certain site investigations be carried out. The purchaser is thus being compelled to complete, notwithstanding that there is uncertainty as to what can be developed and at what cost. This difficulty can be avoided by providing that the trigger for the conditional contract is the grant of a 'satisfactory planning permission' and that a permission shall be deemed not to be 'satisfactory' if subject to certain onerous conditions.

LEASEHOLDERS

Leaseholders as 'owners'

14.15 Some statutory provisions create criminal and civil liability for 'owners' (eg s 79 of the EPA) and it is arguable that tenants often fall within that expression. The appropriate starting point is to look to the relevant interpretation section in the statute.

The Town and Country Planning Act 1990 (TCPA) generally defines 'owner' as being the person for the time being entitled to receive the rack-rent: s 336. If a lessee is able to create a subtenancy, then the subtenant is normally an owner for the purposes of the TCPA. The same definition occurs in s 343 of the Public Health Act 1936. Unfortunately, the word is often left undefined. Neither the Control of Pollution Act 1974 (COPA) nor the EPA do so. This is surprising and a recipe for uncertainty.

In the absence of guidance from a statutory interpretation provision, the question will probably be resolved on the basis of control. This follows from consideration of cases decided in the USA under the Superfund legislation, especially the *Fleet Factors* case: see **14.4**.

Leaseholders: apportioning liability

14.16 Traditional leasehold covenants are not designed to cope with modern regulatory practices, with the result that there can be uncertainty as to liability between landlord and tenant for the costs of remedial measures. The tenant will often accept liability for all impositions or outgoings charged on the premises during the currency of the lease. Such an obligation will, in the absence of a landlord's repairing covenant, make the tenant liable for remedial costs even in respect of a defect in the property when he took the lease: *Foulger v Arding* [1902] 1 KB 700.

14.17 In *Howe v Botwood* [1913] 2 KB 387, the tenant was obliged to meet 'all outgoings imposed on the landlord in respect of the premises'; however, the landlord was also required to keep the exterior of the building in repair. An abatement notice was served requiring that certain works be carried out on external drains, and the High Court held that the landlord was responsible.

If the matter of potential environmental liabilities is left open between landlord and tenant, then this could give rise to problems in the future. The parties would be well advised to try to set out their respective responsibilities in the lease or tenancy agreement. The prudent tenant will seek to exclude all liability for any breaches of environmental laws arising from events occurring outside the period of the tenancy and require the landlord to provide an indemnity for any liabilities arising from such events. If the landlord remains liable for certain events, then he must reserve a right of entry to the land to enable him to carry out remedial works.

It is important that any division of responsibility between landlord and tenant has regard to the way in which regulatory bodies discharge their duties. The regulatory body may, as an alternative to, or preliminary to, formal action, serve an informal and non-statutory 'intimation' notice or letter indicating that, unless specified remedial works are carried out, then formal action may be commenced. Given that the threatened formal action may often involve criminal sanctions, the prudent recipient will often comply with the informal intimation notice or letter.

14.18 In *Harris v Hickman* [1904] 1 KB 563, the landlords expended money in order to comply with an intimation notice but failed to recover the money from the tenants, who were liable for all 'outgoings', because they (the landlords) were deemed to have acted voluntarily instead of under a statutory obligation. If the tenant is made liable for the cost of remedial works, then the relevant provision in the lease should make it clear whether such liability arises only on the service of a formal abatement notice or, alternatively, that it can be triggered by a bona fide non-statutory intimation notice or letter.

Leaseholders: managing the tenancy

14.19 It is not, nowadays, always sufficient for landlord and tenant to rely on covenants which merely apportion liability. If the tenant's operations could generate contamination or pollution, then the prudent landlord should require that the tenant follow a prescribed scheme of environmental audit and management throughout the tenancy – for example, BS 7750 or the EC Eco-Management and Audit scheme. Failure to comply with the scheme would then be a ground for service of a notice under s 146 of the Law of Property Act 1925 and/or give the landlord right of entry to remedy any contamination or pollution caused by the tenant.

CONTAMINATED LAND AND RECEIVERS

14.20 As a general rule, a receiver is the agent of the borrower not the lender. He is appointed for the protection of property either by the court or under a mortgage or debenture.

In *Bacup Corporation v Smith* (1890) 44 Ch D 395, it was held that a court-appointed receiver of rents of a house was not an 'owner' within the meaning of the Public Health Acts, because he did not receive rent either on his own account or as agent or trustee for any other person: see also *Ratford v Northavon District Council* [19878] QB 357. Unless he exceeds his authority or there is a defect in his appointment, a receiver will not normally be liable if anything is wrong with the site – his liabilities are treated as the liability of his principal, the borrower. There are, however, a number of concerns.

14.21 In *Meigh v Wickenden* [1942] 2 All ER 68, a receiver and manager was found guilty under the Factories Act 1937 for inadequately guarded machinery in a factory which the receiver was continuing to run. The duty to provide the guards was placed on the 'occupier' and the receiver's activities fitted that description. In *Lord Advocate v Aero Technologies Ltd* 1991 SLT 134, the Scottish Outer House held that, prima facie, receivers could be 'occupiers' for the purposes of the Explosives Act 1875, because, among other things, they had acted as managers of an explosives factory. It will be recalled that, in certain circumstances, s 80 of the EPA makes the occupier of land liable for statutory nuisance. The position of receivers in relation to criminal offences involving 'causing or knowingly permitting' (eg s 85 of the Water Resources Act 1991) is ambiguous. Many commentators have opined that receivers may become personally liable under such provisions.

The wise receiver should, therefore, seek appropriate indemnities from his principal; however, there is a caveat in relation to indemnities. This is that an indemnity in respect of a criminal penalty is not, as a matter of public policy, enforceable.

PURCHASE BY SHARE ACQUISITIONS

14.22 It is not the purpose of this book to deal with share acquisitions; however, it must be mentioned that a purchase by share acquisition can make the purchaser liable for matters which would not go with a mere asset purchase.

It will be seen from Chapters 5 and 6 that the original polluter can be subject to civil and criminal liabilities which do not attach to his land or assets. They would not, therefore, transfer to a later purchaser. But a share purchaser effectively steps into the shoes of the original owner of the company acquired, because the legal entity comprised in the company remains the same throughout. It would be unfortunate if he stepped straight into a civil or criminal action.

'COMFORT UNDERTAKINGS'

14.23 Having looked at some typical problems relating to primary responsibilities for risk, it is now necessary to consider ways of redirecting liabilities. Whilst a party to a transaction may, as a matter of general law, be directly liable to a regulatory body for the costs of remedial action, that does not necessarily mean that he has to foot the bill. It may be the case that the terms of the original transaction placed a third party under an obligation to reimburse to him the costs of such remedial action.

The time-honoured lawyer's devices designed to secure this end are warranties, indemnities, guarantees and insurance. For ease of exposition, these have been placed under the generic title 'comfort undertakings'; however, it is stressed that this is an expression of the author's own making and will create puzzlement if used outside the context of this book. It was suggested to the author that the phrase '*binding* comfort undertaking' be used to distinguish these undertakings from the non-binding 'comfort letters' which often accompany commercial transactions; however, it is felt that the word 'binding' creates a sense of certainty which might not be borne out in practice. Each such comfort undertaking has, however, its own characteristics and these will be described below – but please note that these descriptions are particular to contaminated land and are not intended to be of universal application.

A number of illustrative comfort undertakings have been set out in Appendix 1. These model documents are not designed to be precedents, but are intended to show how the principles discussed in this chapter might look in practice – albeit in a simplified form.

Warranties

14.24 A 'warranty' is a contractual term that the site is as described by the vendor. Model 2 in Appendix 1 outlines some typical warranties and it will be seen that they cover a variety of matters including potential claims for pollution, compliance with statutory regimes and any abatement or restoration works in relation to the site. A warranty may be *absolute* in that it states conclusively that a given state of affairs does or does not exist: see Model 2 Warranty 1, or it may be *conditional* in that it covers only certain matters or periods of time or is stated to be 'only so far as the Vendor is aware': see Model 2 Warranty 2.

It is important to observe that the normal measure of damages for breach of a warranty is the difference between the current value of the land and its value with proper disclosure of the information in the warranty. This might not be sufficient where the costs of remedying the effects of pollution from a site exceed its value.

Indemnities

14.25 An 'indemnity' is an undertaking that costs, expenses or liabilities incurred by a named person will be reimbursed by another in certain defined circumstances: see Appendix 1 Model 3. For example, X may indemnify Y for all Y's costs or expenses in the event that any abatement notice under s 79 of the EPA is served on Y. If Y is then served with such a notice, X will, in theory, have to reimburse all of Y's costs and expenses including abatement costs – and these may well exceed the value of the site itself. Unlike a warranty, there is no in-built limit on the undertaker's liability under the indemnity.

One often hears of the phrases 'cross-indemnity' or 'counter-indemnity' as though they are special devices; however, this is a matter of terminology, rather than substance, arising where both parties are giving indemnities to each other. Thus, the vendor may give an 'indemnity' on one matter, with his purchaser giving a 'cross-indemnity' on another.

Guarantees

14.26 A 'guarantee' is a contractual bond by a third party supporting a contractual commitment by one of the principal parties to the transaction. Taking the above example one step forward, Y may not be sure that X can afford to cover the indemnity in respect of statutory nuisance, therefore he might require that Z undertake to discharge the indemnity if X cannot – if Z accepts the liability then he will be X's guarantor.

Insurance

14.27 'Insurance' is a form of indemnity whereby, for a premium, the insurer undertakes to pay specified costs and expenses when a specified event takes place. Typical public liability insurance covers civil liability in respect of the death of or injury to any person who is not an employee of the insured and/or damage to that person's property. Since 1991 most UK public liability policies specifically exclude liability based on gradual or past pollution – and, where cover is given for pollution claims, this is confined to sudden and fortuitous incidents which occur, in their entirety, during the period of insurance. The exclusion is intended to prevent coverage of the typical pollution incident where the on-site contamination is historical and its release into neighbouring land or environmental media is over a period of time. Many modern policies also place limits on the making of claims after the expiry of the policy.

14.28 If a modern insurance policy does not cover an incident, then it is worth looking at old policies. Most historical public liability policies cover events occurring during the currency of the policy, regardless of when the claim is made. Thus, one can find that contamination occurring, say, 20 years ago is covered by a policy which has long expired. This situation is, of course, subject to proof, by the insured, that the contamination occurred during the life of the policy – and this task is by no means an easy one: see *Kelly v Norwich Fire Insurance Ltd* [1989] 1 WLR 129.

Most policies also exclude 'own site' damage, which means that the insured cannot claim for remedial works on his own site. The 'own site' exclusion prevents coverage in the not infrequent situation where a regulating body requires that the pollution is remedied at source – often at considerable cost to the landowner.

Furthermore, it is a matter of interpretation of each policy as to whether or not it covers any claims at all in respect of the requirements of a regulatory body – regardless of whether the pollution is sudden and accidental or not.

For example, the standard Lloyd's policy covers 'such sums as the assured shall become legally liable to pay in respect of claims made against the assured arising from ... injury ... to persons or damage to property'. Not all pollution involves damage to property or injury to persons; a regulating body may be acting in respect of pollution of the unowned environment.

14.29 The insurance market has reacted to the perceived growing importance of environmental claims by providing specially designed environmental insurance. There are a variety of policies available including 'environmental impairment liability insurance', 'environmental remediation insurance' and 'own-site clean-up insurance'.

These policies are designed to cover unexpected pollution incidents and, as such, the insurer will usually require that an *environmental assessment* is carried out before accepting the risk. If the insured is carrying out operations on the site which may, potentially, cause contamination, then the insurer may insist on periodical *environmental audits* covering both the condition of the site and the operator's environmental management regime. These preconditions can be extremely expensive and, of course, the premium will be commensurate with the risk.

In passing, it is noted that many *professional indemnity* policies now exclude liability for negligence in respect of pollution. The prudent landowner or developer might therefore be wise to choose consultants who, where appropriate, have professional indemnity cover for pollution.

Conditional comfort undertakings

14.30 If one party to a transaction obtains an absolute unconditional comfort undertaking, then he is doing well – but, in the vast majority of cases, the undertaking will be conditional and subject to a number of limitations.

Indemnities, guarantees and insurances are not subject to any theoretical maximum limit on the undertaker's financial liability and, as such, he will often seek to impose such a limit. The purchase price of the site will often form the ceiling. The undertaker may also seek a threshold minimum below which no claims may be made. The prudent comfort undertaker may seek *time-limits* on his liabilities for both past and future liability. One possibility is staged time-limits whereby different time-limits apply to different types of claims. Another possibility is the setting of time-limits by reference to specified events, for example the carrying out of post-transaction site investigations or the grant of specified licences or permissions.

Limitations may be placed on *types of risk*. For example, the comfort undertaking may be limited by reference to a certain type of pollution only or pollution from a specified source only. It could also be limited by reference to the identity of the polluter. The undertaker might be prepared to accept liability arising from *his* erstwhile activities on the site, but not the actions of others. One example is the contaminant put into the site by the vendor, yet later disturbed and released into neighbouring land by the purchaser's development – it may be that the vendor is not prepared to accept that risk.

Drawbacks and problems

14.31 It is necessary to state, at the outset, the apparently obvious: namely, that there is no point in taking an undertaking from a man of straw. It is essential that the credit worthiness of the comfort undertaker is clearly established at the outset. This means making bankruptcy and other searches and, with a corporate undertaker within a group of companies, obtaining the undertaking from a parent rather than subsidiary company.

If there are doubts as to the undertaker's credit worthiness, then one solution is for him to place a sum of money in an 'escrow' account to be paid out in the event of a claim. An 'escrow' account is a bank account in the joint names of the undertaker and the beneficiary of the undertaking; the balance of monies thus held will be repayable to the undertaker at a defined time or by defined dates. The obvious drawback of an escrow account, for the undertaker, is that it ties up his money. Certainly, one cannot see the regular use of escrow accounts to underpin long-term comfort undertakings.

14.32 If the comfort undertaking is a warranty or indemnity, then it might be necessary for the undertaker to provide a guarantee. If the comfort undertaker is not a man of straw, then he will probably be fond of his money and will not part with it until all elements of the claim have been established. Ambiguities in the wording of the undertaking may lead to arguments and litigation.

14.33 In the field of insurance it is notable that a very successful series of legal conferences has been one entitled 'Denying Coverage in Insurance Claims' – the promotional leaflet for the June 1994 conference promising to, among other things, 'Target and clarify the latest developments and legal twists in the avoidance of liability'.

14.34 The beneficiary of the undertaking must, at the outset, prove that the undertaking has been triggered. That will mean (a) establishing the physical nature of the claim, and (b) establishing that this gives rise to liability within the undertaking. Proving that the site is contaminated or giving rise to pollution is not, in the absence of clear standards (see Chapters 2 and 13), straightforward.

If neighbouring land is polluted, and there is more than just one contaminated site in the locality, then it may be very difficult to establish exactly the contribution of each site to the sum total of pollution. If the undertaking distinguishes between contamination created by different persons (eg vendor as opposed to purchaser or vice versa), then it may be very difficult or impossible to decide, some time after the event, the identity of the polluter. There will be similar problems where the undertaking limits the undertaker's liabilities to a specified period of time only.

If (a) is satisfied, then establishing both liability and the measure of the payment ('quantum') will depend on close analysis of the wording of the undertaking and, with quantum, an expert opinion as to the damage, the remedy and the costs of the remedy. All involve the potential for considerable argument. Certainly, dispute resolution mechanisms can help to ameliorate these problems; however, there will always be a delay in going through such a procedure.

14.35 Since, in theory, it is his money which is going to be spent, the person giving the comfort undertaking will expect to become involved in negotiating with the regulatory body or injured party. Hence, all properly drafted comfort undertakings will include an agreed claims procedure. The procedure will allow for the undertaker's participation in all relevant negotiations and, especially with an insurance policy, may pass all powers of negotiation to the undertaker, including the conduct of any litigation. There may be an inherent conflict of interest in this situation, because what may suit the undertaker may be completely wrong for the business relying on the undertaking. For example, the business may want to limit adverse publicity yet that may be irrelevant to the comfort undertaker.

Another problem in allowing the undertaker conduct of the claim is that regulatory bodies often rely on their criminal powers to secure compliance: thus one could have a situation where the regulatory body's claim is based on a mixture of criminal and civil liability, yet the undertaker is only interested in the civil aspect. One can imagine cases where intransigence over civil liability by the undertaker might trigger criminal charges against the polluter relying on the undertaking, which charges might have been avoided by a less confrontational approach.

The undertaking will probably try for the cheapest solution to the problem; however, a cheap ad hoc solution might not be in the best long-term interests of the landowner. As such, it is necessary that all parties to the undertaking participate in settling the solution – it may be that the person relying on the undertaking will be entitled to 'top-up' the costs of the solution to provide remedial works which give a better long-term result for him.

14.36 Another little discussed limitation is that a comfort undertaking purporting to cover criminal liability cannot, for reasons of public policy, be enforced in any court. An unenforceable indemnity is little comfort to the company chairman who has just been fined £15,000. Similarly, an undertaking will not necessarily meet the costs of complying with an injunction. Civil actions often involve claims for both damages and injunctions and it is often the case that the injunction has the greatest long-term effect on the conduct of the business to which it applies.

Comfort undertakings are not, necessarily, an absolute alternative to expensive site audits. The prudent undertaker may require that an environmental audit be carried out before he gives the undertaking and then require that periodic environmental audits be carried out in respect of the site and the management of on-site activities.

Finally, all such comfort undertakings have a calculable cost. The calculation is difficult with warranties and indemnities but, in theory, it is reflected in the price. One has to pay a premium for an insurance policy which reflects the risk assumed, and guarantors often require a similar payment.

The drafting of undertakings

14.37 Some of the drawbacks discussed above cannot be avoided completely but can be ameliorated by good draughtsmanship. Others are pure drafting problems which can be avoided by careful draughtsmanship. The key drafting problem is deciding exactly *when* the undertaking will be triggered. Standards (or lack of them) pose a significant problem and, in many cases, it may be appropriate for the undertaking to set its own standards.

In many cases, the most sensible undertaking is one which is triggered *before* contamination becomes a source of concern to a regulatory body, or causes injury to a neighbour or an on-site problem. It is difficult, however, to see how such an undertaking could operate without prescribing its own standards – by definition it would have to operate before the regulatory regime or common law comes into play.

14.38 If the undertaking is triggered upon intervention by a regulatory body, then it is necessary to specify exactly when the undertaker's liability may arise. Save in cases of emergency, almost all regulatory bodies give advance warning of their concern – often by way of an 'intimation letter' indicating that formal action may be taken if the recipient does not execute certain abatement and remedial actions. A sensible businessman would act on the basis of an intimation letter in order to avoid additional civil and criminal liabilities and, probably, to keep adverse publicity to a minimum. Consideration, therefore, needs to be given to making a formal bona fide intimation letter from a regulatory body the trigger for the undertaker's liability.

If the liability arises from injury to a neighbour or another party, then the parties to the undertaking have to decide whether the triggering event will be the issue of a writ or court judgment (with all the inherent legal costs) or the receipt of a bona fide 'letter before the action'. If the trigger event is the issue of the writ or the court's final judgment, then the undertaker will expect to be advised of the receipt of any letter before action, or other indication of a pending action, and then empowered to agree a settlement figure out of court, that settlement figure then being the measure of liability.

14.39 Undertakings, especially warranties, often refer to compliance, or part compliance, with 'environmental laws' and then seek to define that phrase by reference to the statutes which are of concern to the parties. It is, however, important to remember that the regulatory bodies tend to proceed by reference to circulars and codes of practice issued by governmental departments – indeed some ministerial documents are given special status by statute: for example, see s 34(10) of the EPA. As such, consideration needs to be given to including compliance with such circulars and codes of practice as a term, or pre-requisite, of the undertaking. There are draft definitions for environmental laws and environmental codes in Appendix 1 Model 5. The parties will also need to consider whether such definitions will include reference to statutes which, in whole or in part, are not in force at the time of the transaction.

Chapter 15

ACCESS TO ENVIRONMENTAL INFORMATION

INTRODUCTION

15.1 The UK is, traditionally, conservative in giving members of the public access to information; however, there are a number of sources of environmental information available and they will be considered in this chapter. We will begin by discussing the debâcle over registers of contaminated land uses because, aside from needing to know the outcome of the debate, one learns from it something about the tensions which surround the publication of information. We will then look at particular sources of information.

REGISTERS OF CONTAMINATIVE LAND USES

15.2 Section 143 of the Environmental Protection Act 1990 (EPA) states that it shall be the duty of the local authority, as respects land in its area subject to contamination, to maintain a register in a form prescribed by the Secretary of State for the Environment. Section 143 was due to be brought into force by regulations produced by the Secretary of State but is now repealed by the Environment Act 1995.

There was considerable concern as to the way in which local authorities were going to assess whether or not land should be included on the register. It seems clear that the criteria for inclusion on registers would have depended on the historical uses of land. Thus, if land had been used in the past for oil refinement or chemical works or the like, then it would probably have to be included on the register despite the absence of direct evidence that any of these activities actually contaminated the land.

Not surprisingly, those acting on behalf of landowners were very critical because, they argued, there would be a major blight upon the value of any land placed on a register. On 24 March 1993, the then Environment Secretary, Mr Michael Howard announced that the Government was abandoning its proposals for the registers. He stated in Parliament that the Government remains:

> 'determined to ensure that land contamination does not give rise to unacceptable risks to health and safety, and also to ensure that vacant urban and industrial land can be put back into good use wherever practicable. The responses to our consultation paper showed that the proposed registers would have made this objective more difficult by reducing confidence in the value of sites placed on the register, thereby exacerbating blight without giving clear indication on how sites could be brought back into good condition and confidence restored.'

The DoE published a consultation paper in March 1994 entitled *Paying for Our Past* and references are made to that paper at relevant places in this book. The consultation paper argued that the private sector has shown itself capable of fulfilling some of the intended purposes of the registers by the use of site-assessment techniques. It also observed that some local authorities have collated material which would have been required under s 143 of the EPA. Having observed that there are some problems in the way the market handles this type of information, the paper indicated that:

> 'It appears that Section 143 of the 1990 Act may not be necessary to ensure an efficient market in land and to ensure that contamination is put right.'

The paper makes no attempt to comprehensively analyse the provision of environmental information in the UK and to assess how the current arrangements can be improved. No doubt this is due to the review committee's terms of reference, but, given the origins of the review, it is very surprising.

THE ENVIRONMENTAL INFORMATION REGULATIONS 1992

15.3 EC Directive 90/313 on freedom of access to information on the environment requires public access to environmental information held by public bodies. The directive has been transposed into UK law by way of the Environmental Information Regulations 1992 (SI 1992/3240), which came into force on 31 December 1992.

15.4 The Directive and regulations provide for a wide range of information on environmental matters held by public authorities to be made available to individuals and organisations. It is inevitable that the regulations will be read in the context provided by the Directive and that, where there are disparities, questions will be raised about whether the Directive is directly enforceable (see **4.27** on direct enforceability). Regardless of direct enforceability, any ambiguities in the regulations will be construed in accordance with the purpose of the directive: see *Marleasing SA v La Comercial Internacional de Alimentacion SA* [1992] 1 CMLR 305 (see **4.27**).

15.5 There have already been criticisms that the new regulations do not conform to the Directive. For example, the directive states that the public must have access to environmental information which is 'available' to a public authority; however, the regulations use the word 'accessible'. Arguably, the two do not have the same meaning. Another criticism has been made that the regulations do not transpose the requirement in the directive that Member States 'shall define the practical arrangements' by which environmental information is 'effectively made available'. The regulations simply impose on the bodies subject to them a duty to make arrangements to comply with the Directive. The Government takes the view that the number of bodies which could be subject to the regulations is so large, and their circumstances so diverse, that it would be impossible to come up with practical arrangements for all of them. On 27 June 1994, the *Independent* reported that the European Commission had formally notified the UK Government about alleged failures in implementation of the directive, including the absence of an effective right of appeal.

15.6 The regulations apply generally to any information which 'relates to the environment' and is held by a 'relevant person': see reg 2(1)(a) and (b). They do not apply to any information or record which is made accessible to the public by any other statutory provision or which is held for judicial or legislative functions: see reg 2(1)(b) and (*c*).

By reg 2(2), information 'relates to the environment' if it relates to the state of water or air, flora or fauna, or any soil, natural site or other land and includes any activities or measures (including noise or any other nuisance) which will or are likely to adversely affect any of them or are designed to protect any of them. Some commentators have opined that, somewhat surprisingly, the regulations may not cover information relating to the state of human beings, bacteria and micro-bacteria (because they may not be 'fauna'): see *Birtles* (1993) JPL 615 at p 616. As such, information about the impact on human conditions after disasters such as the Camelford water pollution incident might not be covered.

15.7 This apparent conundrum may need to be resolved by the courts; it is not helped by the Directive, which – in this particular – is faithfully reproduced in the regulations. Arguably, 'fauna' could be given an extended meaning to cover human beings or, with greater difficulty, bacteria or micro-bacteria. In practice, information relating to unlisted media and subjects may, in any event, be covered by information which has to be made public in relation to others.

Regulation 2(3)(a) defines 'relevant person' subject to the regulations as including all such ministers of the Crown, government departments, local authorities and other persons carrying out functions of public administration at a national, regional or local level as, for the purposes of or in connection with their functions, have responsibilities in relation to the environment. The definition also covers any body with public responsibilities for the environment which is under the control of a person mentioned in reg 2(3)(a): see reg 2(3)(b). The DoE consultation paper which preceded the regulations included a list of 211 bodies affected by the regulations and they are listed, for guidance, in Appendix 2(4) of this book.

In *Griffin v South West Water Ltd* (1994) 236 ENDS Report p 38 Blackburn J ruled that the privatised South West Water Company was an 'emanation of the State' and could be subject to directly enforceable EC legislation. This was a case brought in respect of proposed redundancies; however it may possibly be cited in connected with Directive 90/313 thereby bringing the water companies, British Gas and British Coal within its remit. Information held for judicial or legislative purposes is not environmental information (see reg 2(1)(b)), therefore the courts and those drafting primary and secondary legislation (including ministers acting in that capacity) are excluded.

15.8 Those exercising their right to information under the regulations do not need to prove any particular interest, but they may be required to state the purpose for which the information is requested and to pay the costs of providing the information. The right only applies to information in 'an accessible form'; but it is not clear what that phrase means. Relevant information includes anything contained in registers, reports and returns as well as computer records and other non-documentary records: reg 2(4). The body providing the information must respond to every request as soon as possible and, in any event, within 2 months: reg 3(2).

A request may be refused if it is 'manifestly unreasonable or is formulated in too general a manner': reg 3(3). Such refusal must be in writing and specify the reasons for the refusal: reg 3(2)(c). The problem for the person refused information, or provided with inadequate information, is that there is no effective system of enforcement or appeal. He is left with either the massive expense of judicial review in the High Court or the uncertainty of administrative action through his MP or local councillor or (if the refusal can be described as 'maladministration') through the Parliamentary or Local Commissioner for Administration.

15.9 Clearly, the preservation of commercial and private confidentiality is of considerable concern and reg 4 provides two categories of confidential information, namely either where requests *must* be refused or where they *may* be refused.

By reg 4(3) a request for information *must* be refused where that information:

(a) is capable of being treated as confidential and its disclosure would breach any law or agreement;
(b) is personal and was originally supplied by an individual who has not given his consent to its disclosure;
(c) was originally supplied by a person who was not under any legal obligation to supply it and has not consented to its disclosure; or
(d) would, if released, increase the likelihood of damage to the environment (eg information about nesting sites, rare habitats, etc).

The regulations have been criticised in relation to item (a) because they appear to be wider than the corresponding provision in the Directive, which stipulates that a Member State may provide that a request for information may be refused 'where it affects ... commercial and industrial confidentiality'. This issue could be tested in the courts in a case involving the Crown Estate: see the *Guardian*, 25 May 1994.

By reg 4(1),(2) and (5), a request for information *may* be refused where the information:

(a) relates to international relations, national defence or public security;
(b) relates to anything which is, or has been, the subject of any legal or other

proceedings (whether actual or proposed) including public inquiries and hearings by appointed persons;
(c) relates to the confidential deliberations of any relevant person defined by reg 2(3) or to any internal communication by a body corporate or other undertaking or organisation;
(d) is contained in any document or record which is still in the course of completion; or
(e) affects matters to which any commercial or industrial confidentiality attaches or any intellectual property.

LOCAL AUTHORITIES

15.10 The Local Government (Access to Information) Act 1985 amended the Local Government Act 1972 to create public rights of access to local authority information. Subject to certain limitations, these provisions allow public access:

(a) to meetings of the full council, committees and sub-committees;
(b) to agendas, minutes and reports for such meetings; and
(c) to background papers for such meetings.

Schedule 12A to the 1972 Act sets out the exemptions to the right of access, viz:

(a) information relating to the council's employees or persons receiving financial benefits or accommodation from it;
(b) information relating to the care of any child;
(c) information relating to the financial or business affairs of any person;
(d) contract and tender information;
(e) information relating to the prevention, investigation or prosecution of a crime; and
(f) information which would reveal that the authority proposes to serve or make a statutory notice or order if the revelation of that information would defeat the purposes of the notice or order.

It is also worth remembering that, whilst the s 143 registers are now of historical interest only, many local authorities were engaged in work preparatory to setting them up or, indeed, have created their own informal registers – as such it is worth checking whether the local authority for the relevant site has this information available.

Specifically in relation to contanminated land, the Environment Act 1995 now provides that local authorities and the Environment Agency will have to keep statutory registers in relation to the identification of contaminated land, the designation of special sites and certain measures which have been taken in respect of them: see Chapter 16.

Borough and district councils maintain local land charges registers under the Local Land Charges Act 1975, which may contain details of planning obligations relating to environmental matters.

LICENSING REGIMES

15.11 Most statutory regimes for the regulation of activities affecting the environment involve the holding of registers which are open to the public. The information that can be found typically includes:

(a) details of applications and grants;
(b) consultee responses;
(c) applications for transfer, variation or surrender of licences or authorisations;
(d) details of any appeals; and
(e) details of any convictions of the licence/authorisation holder for breaches of the statutory regime.

Register information can be obtained from the following bodies in relation to the following regimes:

(a) Waste regulation authorities/Waste collection authorities – waste management licences;
(b) Waste regulation authorities – registration of waste carriers;
(c) Her Majesty's Inspectorate of Pollution (HMIP) – integrated pollution control;
(d) Local authorities – local authority air pollution control;
(e) HMIP and local authorities – information under the Alkali etc Works Act 1906;
(f) The National Rivers Authority – discharge and trade effluent consents;
(g) HMIP – radioactive substances;
(h) Health and Safety Executive and local authorities – health and safety at work;
(i) Local authorities – town and country planning;
(j) HMIP – genetically modified organisms; and
(k) HMIP – the Chemical Release Inventory.

More detailed advice can be obtained from *Documentary Research on Industrial Sites* (Contaminated Land Report No 3) published by the DoE.

COMMERCIAL CONFIDENTIALITY

15.12 Many regimes exclude public access to information about trade secrets. For example, commercial confidentiality exemptions may be claimed, by affected persons, in the following cases:

(a) waste management licensing: s 66 of the EPA;
(b) NRA discharge consents: Sch 10 to the Water Resources Act 1991;
(c) integrated pollution control and local authority air pollution control: s 22 of the EPA;
(d) radioactive substances: s 13A of the Radioactive Substances Act 1960; and
(e) registrations under the Alkali etc Works Act 1906.

In all cases, the exemption will arise only if sought by the affected business in the manner prescribed by the relevant legislation. In a number of cases there is a right of appeal against first-instance refusal to grant exemption. It is up to the person claiming the exemption to demonstrate that the relevant information is commercially confidential and that its publication would unreasonably prejudice that person's business interests.

Chapter 16

THE ENVIRONMENT ACT 1995

INTRODUCTION

16.1 The Environment Act 1995 received the Royal Assent on 19 July 1995, although it will be brought into force in stages. If the Control of Pollution Act 1974 and the Environmental Protection Act 1990 (EPA) are anything to go by, the bringing into force of specific provisions could take some time or, indeed, may never happen at all: see **3.5–3.7**.

It is notable that the Act repeals s 143 (on registers of contaminative uses) and s 61 (on closed landfills) of the EPA despite the fact that neither have ever been brought into force. Part II of the EPA (the new regime for controlled waste management) eventually came into force in May 1994 and, even now, there are still gaps to be filled; s 158 of the EPA (on executive liability) is still not in force. The examples are legion. It is to be hoped that implementation of the Environment Act 1995 will be neither as haphazard nor as uncertain.

The Act is wide ranging and addresses matters such as setting up the Environment Agency, abolishing the National Rivers Authority and transferring its powers, and certain powers of Her Majesty's Inspectorate of Pollution and the waste regulation authorities, to the Environment Agency. It also addresses the regulation of pollution from contaminated land – that topic is the focus of this chapter.

Part II of the Environment Act 1995 adds a new Part IIA to the EPA and amends the law relating to abandoned mines. Part II is not yet fully in force and will come into force when the Secretary of State produces the necessary regulations: s 125(3).

16.2 The first commencement order was passed on 21 July 1995, namely, the Environment Act 1995 (Commencement Order No 1) Order 1995 (SI 1995/1983). The order brought s 1 of the 1995 Act into force on 28 July 1995, thereby bringing the Environment Agency into life and, from the same date, bringing into force a number of provisions relating to setting up the Agency. The order, inter alia, brings s 57 (contaminated land), s 58 (abandoned mines) and parts of Sch 22 (anti-pollution works and operations) into force on 21 September 1995 but only so far as to enable the Secretary of Stare to issue the ministerial guidance/directions and regulations which are the necessary preliminary to the operation of those provisions.

All references in this chapter to Part IIA and section numbers will, unless indicated otherwise, be references to the EPA as amended.

THE ROLE OF MINISTERIAL GUIDANCE

16.3 The operation of Part IIA of the EPA is heavily dependent upon guidance from the Secretary of State. This guidance operates at two different levels. First, a number of provisions require that the enforcing body 'have regard to' such guidance (this type of guidance will be called 'advisory guidance' in this chapter). Secondly, a number of other provisions require that the enforcing authority act 'in accordance with' such guidance (this type of guidance will be called 'mandatory guidance' in this chapter).

As to the first type of guidance, advisory guidance, the enforcing authority must take such guidance into account as a material consideration in reaching the relevant decisions; however, the requirement to 'have regard to' something does not, as a matter of law, mean that the determining body has to slavishly adhere to it: *Enfield London Borough Council v Secretary of State for the Environment* [1975] JPL 155 and *Chelmsford Borough Council v Secretary of State for the Environment* [1985] JPL 554. The enforcing authority may depart from such guidance provided that it has fully considered it, taken it into account and given it proper weight.

The second type of guidance, mandatory guidance, is more problematic. The requirement that the enforcing authority act 'in accordance with' guidance from the Secretary of State is clearly obligatory and proceeds on the basis that the enforcing authority will have no leeway to depart from that advice. Failure to follow that advice properly would make the relevant actions ultra vires and could invalidate any dependent subsequent procedures. This is a point of considerable concern because ministerial guidance is not normally noted for its precision of expression. Ambiguities in the wording of such guidance could be the cause of much uncertainty.

16.4 Section 78YA of the EPA provides certain procedural constraints in relation to the issue of ministerial guidance. Before issuing any guidance under Part IIA (whether advisory guidance or mandatory guidance), the Secretary of State must consult the Environment Agency and such other bodies or persons as he may consider appropriate: s 78YA(1). In relation to mandatory guidance, the draft of such guidance must be laid before each House of Parliament and must not be issued until 40 days have passed: s 78YA(2). If, within that 40-day period, either House resolves that the draft guidance should not be issued, then the Secretary of State cannot issue that guidance: s 78YA(3).

On 10 May 1995, the Secretary of State for the Environment issued draft guidance (dated 5 May 1995) on the definition and identification of contaminated land, with the promise that further guidance would be issued in due course on remediation standards and notices, special sites and 'appropriate persons' (see **16.36**) on whom remediation notices could be served.

IDENTIFICATION OF CONTAMINATED LAND

The statutory basis

16.5 Section 78B(1) provides that every local authority shall cause its area to be inspected from time to time for the purposes of:

(a) identifying contaminated land; and
(b) enabling the authority to decide whether any such land is land which is required to be designated as a special site.

Part IIA does not apply to land which is the subject of a current waste management site licence under Part II of the EPA (see Chapter 8) except where the relevant contamination or pollution is attributable to causes other than breach of the conditions of the licence or the carrying on, in accordance with the conditions of the licence, of any activity authorised by the licence: s 78YB(2).

16.6 'Contaminated land' is defined, by s 78A(2), as any land which appears to the local authority to be in such a condition, by reason of substances in, on or under the land that:

(a) significant harm is being caused or there is a significant possibility of such harm being caused; or,
(b) pollution of controlled waters is being, or is likely to be caused.

Confusingly, this use of the word 'contaminated' (as synonymous with harm or pollution) departs from some previous official usage. For example, in the consultation paper *Public Registers of Land Which May Be Contaminated*, the Department of the Environment stated:

> '3.1 The Government share the view of the Royal Commission on Environmental Pollution that contamination is not synonymous with pollution. The Commission regarded contamination as "a necessary but not sufficient condition for pollution", and also noted that "what today is regarded as no more than contamination may tomorrow be pollution, as a substance accumulates to … harmful concentrations, or as knowledge is gained of hitherto unsuspected effects."
>
> 3.2 The Commission therefore used "contamination" to denote the presence in the environment of certain substances, without making a judgment on whether they are actually harmful. The use of the term in this paper follows this principle.'

This is a very important provision and it is necessary to break it down and analyse it thoroughly.

'Substance' means any natural or artificial substance, whether in solid or liquid form or in the form of gas or vapour: s 78A(9). The remediation notice provisions encompass contaminative substances which are the products of chemical reactions or biological processes: s 78F(9). As to the position in relation to natural substances, see **16.39** below.

'Harm' is further defined as harm to the health of living organisms or other interference with the ecological systems of which they form part and, and in the case of man, includes harm to his property: s 78A(4).

'Pollution of controlled waters' means the entry into controlled waters of any poisonous, noxious or polluting matter or any solid waste matter: s 78A(9). 'Controlled waters' are defined by reference to the Water Resources Act 1991: s 78A(9) and see **7.4–7.6**. The reference to 'solid waste matter' is similar to analogous provisions in the Water Resources Act 1991 and means that inert material which does not cause a chemical or biological reaction in the receiving environment falls within that part of the definition – something which can easily catch out the unwary. For example, if inert demolition materials are placed so close to a watercourse that they are likely to slip in, then the local authority might be entitled to serve a remediation notice.

Where two or more different sites, when considered together, are contaminated land then Part IIA applies to each of these sites, whether or not each different site so qualifies by itself: s 78X(1). If a contaminated site straddles the boundaries of adjacent local authorities, then either authority may treat that land as if it were land situated on its area: s 78X(2).

16.7 Section 78A(5) provides that the questions:

(a) what harm is to be regarded as 'significant';
(b) whether the possibility of significant harm being caused is 'significant'; and
(c) whether pollution of controlled waters is being or likely to be caused.

shall be determined in accordance with mandatory guidance issued for that purpose by the Secretary of State.

Section 78A(6) provides that the guidance may make provision for different degrees of importance to be assigned to or for the disregard of:

(a) different descriptions of living organisms or ecological systems;
(b) different descriptions of places; or
(c) different descriptions of harm to health or property, or other interference.

The guidance may make provision for different degrees of possibility to be regarded as 'significant' (or as not being 'significant') in relation to different descriptions of significant harm.

The definition of contaminated land in the Environment Act imposes a subjective test for contamination by including the words 'which appears to the local authority'. This approach is not particularly conducive to consistency from authority to authority or to certainty of meaning. Presumably, the guidance produced by the Secretary of State will be designed to create consistency of approach. The local authority will, in making its decision, have to comply with the usual requirements of administrative law including the requirement to act 'reasonably': see **3.12–3.13** above.

The practical problems

16.8 The enforcing authorities, and the Secretary of State in formulating guidance, will have to contend with some extremely difficult problems. First, coming up with a precise description of what qualifies as contamination within Part IIA for a particular site may prove extremely elusive. Expert opinions may (except in the most obvious cases) differ as to whether a possibility of harm is 'significant' or pollution of controlled waters is 'likely' in a particular case, and general guidance from central government may not resolve the problem because every site differs in its characteristics and its propensity to emit pollutants. Secondly, identifying the site which is causing a pollution problem may prove exceptionally difficult, and, again, expert opinions may vary. This is especially so with the pollution of controlled waters, where pollution can travel for many miles, often via groundwaters. Finally, this type of investigation can be extremely costly, involving site investigations (eg bore-holes, trial-pits and sophisticated laboratory analysis of samples). As such, the practicalities of this duty to identify contaminated land needs to be considered in light of the matters discussed at Chapters 2 and 13 of this book.

16.9 It is also worth bearing in mind that expert knowledge and opinion on contamination and pollution are constantly being redefined, with the result that land which may be regarded as innocuous one day could be regarded as 'contaminated' the next due to a change in attitude to the substances found on that land. This means that a local authority could conclude that a certain site is not, on current standards, contaminated land for the purposes of Part IIA, yet change its mind in the future as a result of new scientific opinion. Indeed, one could have a situation where a site is cleared to a point which discharges a remediation notice (see **16.21** and **16.28** below) on current standards, yet find that another remediation notice has to be served in the future because the residual levels of contamination are then found to be unacceptable. As such, neither the identification procedure nor the remediation notice procedure can be regarded as 'once and for all'. Potential purchasers should be particularly aware of this point.

16.10 It will be interesting to see how local authorities discharge their statutory duty to inspect their areas. The duty appears to be pro-active in the sense that the authorities ought to engage in some systematic programme of inspection throughout their areas, but it would not be surprising if the authorities merely performed reactively upon the receipt of complaints only. The key question is, of course, the availability of funds – yet no additional funding is proposed.

The draft guidance

16.11 Paragraph 27 of the draft guidance of 5 May 1995 (see **16.4**) gives the somewhat vague advice that:

'In carrying out its duty to inspect its area from time to time for the purpose of identifying contaminated land the authority should deal with contaminated land in an ordered way whilst ensuring vigilance in the identification of actual threats to health or the environment.'

On a more practical note, para 29 states that:

'The authority should prepare an appropriate local strategy for carrying out the inspection of its area which will make effective use of resources and ensure that effort is concentrated first on areas where the chance of discovery of significant harm is greatest.'

The draft guidance makes no attempt to face the cost implications of site investigations and scientific testing.

16.12 In the draft guidance the Secretary of State proposed that, for the purposes of defining contaminated land, the enforcing authorities should disregard any harm or interference other than:

- harm to health of human users or occupiers of the land in its current use or to health of current human users or occupiers of other land;
- harm to or interference with the ecosystems protected under the Wildlife and Countryside Act 1981, EC Directive 79/409/EEC on the Conservation of Wild Birds, or EC Habitats Directive 92/43/EEC, particularly in the light of the authorities' statutory duties under reg 3(4) of the Conservation (Natural Habitats & c) Regulations 1994; and
- harm to property [including livestock and crops], in relation to the present use of the land or other land (para 24).

As to the degree or measure of harm, the draft guidance proposed that the enforcing authorities should disregard any harm or interference other than in the case of:

- harm to human health – death, serious injury or clinical toxicity;
- harm to or interference with ecosystems – a significant change in the functioning of the ecosystems in protected areas, such as sites of special scientific interest, Ramsar sites, or European sites under the Conservation (Natural Habitats & c) Regulations 1994; and
- harm to property – physical damage which is continuing and which cannot be put right without substantial works and, in the case of animals and crops, disease or other physical damage causing loss in value (para 25).

Paragraph 26 of the draft guidance added that:

'Any harm to living organisms or to property or interference with an ecosystem should be disregarded where the harm or interference is an intended and legal result of the addition of substances to the land, for example where harm is caused to a pest by the controlled use of pesticides.'

16.13 The emphasis is on the 'suitable for use' approach and the draft guidance reiterates the Government's policy on this point as set out in the policy document *Framework for Contaminated Land*: see **13.2–13.3**.

The draft guidance proposes that the approach to the determination of the health and environmental effects arising from contaminated land should follow the 'fundamental principles of risk assessment', that is to say 'risk is defined as the combination of the probability, or frequency, of occurrence of a defined hazard and the magnitude of the consequences' (para 35).

The approach taken in the draft guidance is the 'source–pathway–target' approach. In order to illustrate the approach, the draft guidance gives the example of a source–pathway–target relationship where the target is a child and the source is cadmium in soil:

> 'For there to be significant possibilities of a child suffering harm as a result of exposure to cadmium in soil, the source must be accessible through one or more pathways. Possible pathways include the ingestion or inhalation of contaminated soil, skin contact with the soil, or the consumption of contaminated vegetables.
>
> If a child has access in one of these ways to the soil containing cadmium a plausible source–pathway–target relationship has been established. If, however, the soil is all beneath a concrete hard standing, then a plausible relationship is unlikely to be established.'

NOTICE OF IDENTIFICATION

16.14 If a local authority has identified any contaminated land in its area, then s 78B(3) provides that it must give notice to:

(a) the Environment Agency;
(b) the owner of the land;
(c) any person who appears to the authority to be in occupation of the whole or any part of the land; and
(d) each person who appears to the authority to be an 'appropriate person' (see **16.36** below for the definition of 'appropriate person').

There is no requirement to notify the owners of adjoining or adjacent land at this stage (see **16.33** and **16.35**).

16.15 The local authority has to keep a public register of a number of matters relating to contaminated land (see **16.53**); however, the identification notices are not placed on the register. This is a deliberate omission. When the final amendments to the Environment Bill were debated in the House of Commons on 28 June 1995, the Under Secretary of State, Sir Paul Beresford, stated:

> 'The provisions in the Bill relating to registers are set out so that a site is entered on the register only when the regulatory action takes place or when voluntary action takes place or when voluntary action is agreed upon.'

16.16 The Environment Act appears to treat this identification procedure as being some form of administrative procedure internal to the local authority with no input from anyone else, including any person interested in the land or neighbouring landowners, and ending the issue of a notice which is not subject to any form of statutory consultation, review or appeal. No doubt the recipient of a contaminated land identification notice can, in practice, approach the local authority to change its mind – but this does, it is submitted, create an unsatisfactory state of affairs. The issue of such a notice could impact upon the value of the subject land and could create a form of blight. It is unfortunate that the Environment Act does not allow the recipient some form of statutory challenge.

As noted above, s 78B(3)(a) provides that notification needs to be given, inter alia, to the Environment Agency. It would be sensible for any person interested in the subject land, if he is unhappy with the notification, to ask the Environment Agency to dispute the identification. The alternative is for the person aggrieved by the notice to challenge the identification of the site by way of judicial review in the High Court see **3.10–3.18**.

If judicial review is not practicable in the particular case, then another approach might be by way of the Local Commissioner for Administration (see **3.22–3.29**), but the Local Commissioner could intervene only if the identification of the site could be shown to be maladministration. If the local authority has a soundly based opinion on the condition of the site, then it will not be possible to challenge the merits of the identification by way of the Local Commissioner.

SPECIAL SITES

16.17 Part IIA provides that certain types of land which are identified as contaminated land shall be further classified as 'special sites', with the consequence that the responsible enforcing authority will be, for so long as the special site designation applies to the site, the Environment Agency. If at any time it appears to a local authority that any contaminated land in its area might be land which is required to be designated as a special site, the authority shall decide whether or not the land is required to be so designated: s 78C(1)(a). The phrase 'special site' is not comprehensively defined in Part IIA. Section 78C(8) provides that land is required to be designated as a special site if, and only if, it is of a description prescribed in regulations issued by the Secretary of State. Such regulations may make different provision for different areas or localities and may, in particular, describe land by reference to the area or locality in which it is situated: s 78C(9). Section 78C(10) provides that the Secretary of State may, in drafting the regulations, have regard to:

(a) whether land of the description in question appears to him to be land which is likely to be in such a condition, by reason of substances in, on or under the land that—
   (i) serious harm would or might be caused, or
   (ii) serious pollution of controlled waters would be, or would be likely to be, caused; or
(b) whether the Environment Agency is likely to have expertise in dealing with the kind of significant harm, or pollution of controlled waters, by reason of which land of the description in question is contaminated land.

Before making any decision in respect of a proposed special site designation, the local authority must request the advice of the Environment Agency and, in making its decision, have regard to any advice given by the Agency in response to that request: s 78C(3). This means that lobbying the Agency (see **16.16** above) could prove productive. If the Environment Agency consider that any contaminated land ought to be designated as a special site, then the Agency may notify the relevant local authority: s 78C(4). The authority must then decide whether or not the land should be so designated: s 78C(5) and (6).

16.18 If the local authority decides to designate the land as a special site, then it must give notice to 'relevant persons', and the relevant persons are defined, by s 78C(2), as being:

(a) the Environment Agency;
(b) the owner of the land;
(c) any person who appears to the local authority concerned to be in occupation of the whole or any part of the land; and
(d) each person who appears to that authority to be an appropriate person.

16.19 The designation will take effect on the day after whichever of the following events occurs first:

(a) the expiration of 21 days from the day when the notice referred to in **16.17** and **16.18** above was given to the Environment Agency; or
(b) if the Agency gives notification to the local authority in question that it agrees with the decision, the giving of that notification: s 78C(6).

16.20 As with the initial identification of the site as contaminated land, there is no provision for statutory review or appeal in respect of a designation. Again, a person aggrieved by the decision may have to consider reference to the Local Commissioner for Administration or the High Court: see **16.16** above. But the position in relation to lobbying the Environment Agency is stronger because, if the Agency disagrees with the authority's decision, it may serve notice on the authority and the issue will then be referred to and determined by the Secretary of State: s 78D.

REMEDIATION NOTICES

When they may be served

16.21 The enforcing authority in relation to a special site is the Environment Agency, and in relation to other contaminated land is the local authority: s 78A(9). The basic procedure in relation to remediation notices is identical whether the enforcing authority is the local authority or the Environment Agency. If land has been designated a special site or contaminated land, then s 78E(1) provides that the enforcing body must serve each 'appropriate person' with a remediation notice. The phrase 'appropriate person' is discussed at **16.36** below. The procedure to be followed will be prescribed by regulations issued by the Secretary of State: s 78E(1) and (6)(b).

Whilst the duty to serve a remediation notice appears to be immediate and mandatory, this is not necessarily the case. As will be described below, the service of a remediation notice is subject to a consultation stage, constraints as to time of service and an embargo on service if certain conditions apply.

Consultation procedures

16.22 The enforcing body has to consult with various interested persons before issuing a remediation notice (see below) and may, as a result of those consultations, decide that the service of notice is inappropriate (see below), in which case the 'remediation declaration' and/or 'remediation statement' procedures will apply: see **16.26** and **16.27**. The consultation procedures may themselves cause the issue of a remediation notice to be delayed for a considerable period of time.

Save for the special cases mentioned in s 78H(4) (see **16.24**), the consultation procedures have to be carried out before a remediation notice can be served and involves two sets of provisions in two separate sections of Part IIA. First, s 78G(3) provides a consultation requirement in respect of persons other than appropriate persons who are subject to duties arising from remediation notices: see **16.33**. Secondly, s 78H(1) provides that, before serving a remediation notice, the enforcing authority shall reasonably endeavour to consult:

(a) the person on whom the notice is to be served;
(b) the owner of any land to which the notice relates;
(c) any person who appears to that authority to be in occupation of the whole or any part of the land; and
(d) any person of such description as may be prescribed by regulations.

The procedure to be followed may be prescribed by regulations: s 78H(2).

16.23 Neither s 78G(3) nor s 78H(1) provides a prescribed consultation period; presumably the period or periods will be prescribed by regulations. Otherwise, the period or periods actually observed in each case will have to be reasonable in the circumstances of each particular consultee. If the contamination involves a very complex remediation package, perhaps involving numerous parties and regulatory bodies, then the consultation procedures could, reasonably, take months or, in the very extreme case, years.

It is important to observe that the consultation requirements do not mean that a single consultation period will apply to all consultees. No doubt there will be many cases where the picture changes as a result of consultation (indeed it would be a pointless exercise of this were not the case) and new parties may need to be consulted as a result of new information: for example, additional 'appropriate persons' may be found. Each new party who falls within ss 78G(3) or 78H(1) ought to be consulted separately.

Constraints on time of service

16.24 Section 78H(3) provides that a remediation notice cannot, subject to s 78H(4), be served during three time periods:

(a) before 3 months have passed after the service of the notices confirming identification of the contaminated land to appropriate persons (see **16.14** above);
(b) if the local authority decides that the site should be designated as a special site, then within 3 months of the date that the designation notices are served under s 78C(6) or, if the matter is referred to the Secretary of State, within 3 months of the date of his notification under s 78D(4)(b); or
(c) if the Environment Agency notifies the local authority that the land should be a special site (see **16.17**), the period beginning with the day on which the notice is served and ending with the expiration of 3 months beginning with:
  (i) in the case where notice is given under s 78C(6) (ie special site designation: see **16.19**), the day in which that notice is given;
  (ii) in a case where the authority makes a decision not to designate land as a special site under s 78C(5)(b) and the Agency fails to give notice under s 78D(1) that disagrees with that decision (see **16.20**), the day following the expiration of the period of 21 days mentioned in that paragraph; or
  (iii) in a case where the authority makes a decision not to designate land as a special site under s 78C(5)(b), which decision is referred to the Secretary of State under s 78D (see **16.20**), the day on which the Secretary of State gives the notice required by s 78D(4)(b).

Section 78H(4) provides that neither s 78H(1) (see **16.22**) nor s 78H(3) shall preclude the service of a remediation notice in any case where it appears to the enforcing authority that the land in question is in such a condition, by reason of substances in, on or under the land, that there is imminent danger of serious harm or serious pollution of controlled waters being caused.

As to (a), above, it is important to note that s 78H(3) does not provide an automatic minimum consultation period of 3 months following the service of identification notices. As observed above, there is currently no specific prescribed period for consultation. Notwithstanding these comments, one could reasonably expect that the enforcing authority would initiate the consultation procedure at the same time as serving an identification notice (so far as it can at that time), thereby making the 3 month period the minimum consultation period for the recipients of the identification notices.

Embargos on service

16.25 Section 78H(5) provides that a notice may not be served on an appropriate person if and so long as one or more of the following conditions is satisfied:

(a) the authority is satisfied, in consequence of the 'cost-harm' test in s 78E(4) and (5) (see **16.28**) that there is nothing by way of remediation which could be specified in a remediation notice served on that person;
(b) the authority is satisfied that the appropriate things are being, or will be, done by way of remediation without the service of a remediation notice on that person;
(c) it appears to the authority that the person on whom the notice would be served is the authority itself; or
(d) the authority is satisfied that the powers conferred on it by s 78N (see **16.50**) to do what is appropriate by way of remediation are exercisable.

As to (a), the authority may decide that the issue of a notice is not appropriate in the light of the cost-harm test in s 78E(4) or advisory guidance issued by the Secretary of State under s 78E(5). The condition allows the authority to take into account the appropriate person's proposals for remediation and to refrain from service of a remediation notice where it is satisfied that the same results can be achieved by way of a negotiated solution.

The limitations in s 78H(5) are not designed to be a permanent bar to the service of a remediation notice. If it later transpires that the conditions in s 78H(5) cease to apply to an appropriate person, then he must be served with a remediation notice (s 78H(10)), even though the proposed recipient might then, again, be able to contend that the s 78H(5) limitations ought to reapply. In particular, he might bring forward a new or revised package of remediation measures under s 78H(5)(b).

Remediation declarations and remediation statements

16.26 If a remediation notice cannot specify a particular thing by way of remediation because of s 78H(5)(a) (by reference to the 'cost-harm' test (see **16.28**) then a 'remediation declaration' must be published. Section 78H(6) provides that a 'remediation declaration' shall record:

(a) the reasons why the authority would have specified that thing; and
(b) the grounds on which the authority is satisfied that it is precluded from specifying that thing in such a notice.

Nothing in Part IIA indicates how the publication requirement is to be discharged – aside from entering the declaration in the appropriate register: see **16.53**.

16.27 Section 78H(7) provides that where the enforcing authority is precluded, by virtue of s 78H(5)(b), (c) or (d) from serving a remediation notice, then the 'responsible person' shall prepare and publish a 'remediation statement' which shall record:

(a) the things which are being, have been, or are expected to be, done by way of remediation in the particular case;

(b) the name and address of the person who is doing, has done, or is expected to do, each of those things; and
(c) the period within which each of those things is being, or is expected to be, done.

For the purposes of s 78H(7) the 'responsible person' is:

(a) in a case where the condition in s 78H(5)(b) is satisfied, the person who is doing or has done, or who the enforcing authority is satisfied will do, the things there mentioned; or
(b) in a case where the condition in s 78H(5)(c) or (d) is satisfied, the enforcing authority.

As to (a), s 78H(9) provides that, in certain cases of default by a responsible person, the enforcing authority may publish the remediation statement and recover the costs from the person responsible.

What remediation notices can require

16.28 A remediation notice must specify what the recipient is required to do by way of remediation and the periods within which he is required to do each of the things so specified: s 78E(1). Part IIA provides that the form and content of remediation notices may be prescribed by regulations: s 78E(6). However, by s 78E(4), the enforcing authority must specify only remediation actions which it considers reasonable having regard to:

- the cost which is likely to be involved; and
- the seriousness of the harm or pollution of controlled waters in question.

In determining what the recipient is to be required to do by way of remediation, the standard to be achieved and what is to be regarded as reasonable for s 78E(4), the enforcing authority must have regard to any advisory guidance issued by the Environment Agency or the Secretary of State: s 78E(5). As noted at **16.25** above, if the authority takes the view in consequence of s 78E(4) and (5) that a remediation notice ought not to be served at that time, then condition (a) of s 78H(5) will apply and service of a notice will be precluded. The 'remediation declaration' procedure under s 78H(6) will then apply (see **16.26**) but *not* the 'remediation statement' procedure – because s 78H(7) does not apply to condition (a).

Meaning of 'remediation'

16.29 'Remediation' is extensively defined by s 78A(7) and includes the carrying out of works, operations or steps to prevent, minimise, remedy or mitigate the effects of any significant harm or any pollution of controlled waters, or to restore the affected land or waters to their former state. Remediation includes the doing of anything for the purpose of assessing the condition of the contaminated land, any controlled waters affected by the contaminated land or any land adjoining or adjacent to that land. This covers site inspections and surveys, including intrusive investigations such as boreholes and trial-pits, and post-remediation inspections and surveys. The point about access to adjoining or adjacent land is discussed at **16.33** below.

It is possible that the enforcing authority might need detailed survey information before it can require specific works. In that case, it may serve remediation notices merely requiring that certain investigations be carried out – these would then be followed by further remediation notices requiring that specific detailed measures be carried out.

The reference to restoration of affected land or waters 'to their former state' may be a source of future confusion because there may be considerable differences of opinion, first, as to the date or period which ought to be taken as reflecting the pre-contamination state of the land or water, and secondly as to its state at that time.

Costs and benefits

16.30 If the enforcing body is the Environment Agency, then it must, before issuing a remediation notice, take into account the costs which are likely to be incurred and the benefits which are likely to accrue: s 39(1) of the Environment Act. This means that the Agency has to apply two different tests in reaching its decision: first, the 'cost-benefit' test in s 39(1); and secondly, the 'cost-harm' test in s 78E(4) (see **16.28**). It remains to be seen whether or not this will cause confusion in the future.

Avoidance of duplication

16.31 Section 78YB(3) provides that if, in a case falling within s 59(1) or (7) of the EPA (see **8.27–8.32**), the land is contaminated land, or becomes such land, by reason of the deposit of controlled waste, a remediation notice shall not be served in respect of that land by reason of that waste or any consequences of its deposit, if and to the extent that it appears to the enforcing authority that the powers of waste regulation authority (ie the Environment Agency from April 1996) or waste collection authority under that section may be exercised in relation to that waste or the consequences of its deposit.

Section 78YB(4) further provides that no remediation notice shall require a person to do anything the effect of which would be to impede or prevent the making of a discharge in pursuance of a discharge consent given under the Water Resources Act 1991.

Part IIA makes special provision for radioactive contamination in that it does not apply to contamination attributable to such radiation except as provided by regulations issued by the Secretary of State: s 78YC.

Combined sites

16.32 A local authority may serve remediation notices on sites which, when combined, cause, or are likely to cause, harm or pollution of controlled waters even if each site, when considered individually, may not do so: s 78P.

Liabilities of 'non-appropriate persons'

16.33 Part IIA creates certain liabilities and duties for persons who are not appropriate persons, particularly neighbouring landowners. Section 78G(1) provides that the notice may require an appropriate person to do things by way of remediation 'notwithstanding that he is not entitled to do those things'. In such a situation, any person whose consent is required before anything required by a remediation notice may be done shall grant, or join in granting, such rights in relation to any 'relevant land or waters' as will enable the appropriate person to comply with the notice: s 78G(2)

Section 78G(3) provides that the enforcing authority shall reasonably endeavour to consult every person who appears to the authority:

(a) to be the owner or occupier of any of the 'relevant land or waters'; and
(b) to be a person who might be required by s 78G(2) to grant, or join in granting, any rights concerning the rights which that person may be so required to grant.

'Relevant land or waters' are defined as:

(a) the contaminated land in question;
(b) any controlled waters affected by that land; or
(c) any land adjoining or adjacent to that land or those waters (s 78G(7)).

No sanction is prescribed in the event that the non-appropriate person refuses to grant the necessary rights to the appropriate person. Presumably, the provisions could be enforced by injunction, but this does appear to be an odd omission.

16.34 Section 78G(5) provides a somewhat uncertain palliative for the person who grants or joins in granting any rights pursuant to s 78G(2) in the form of a right to claim compensation from the appropriate person. The amount of compensation will be determined in such a manner as may be prescribed by regulations made by the Secretary of State. This provides little comfort if the appropriate person does not have the financial ability to meet such a claim. It is difficult to see how the pursuit of such a compensation claim will be straightforward – save where the appropriate person is very obliging. If he is resistant to payment, then pursuit of a claim could be extremely troublesome. It is surprising that the Environment Act does not provide that the enforcing authority pays the compensation in the first instance and then entitles it to reclaim that money from the appropriate person.

The person granting rights would, therefore, be prudent if he ensures that compensation is covered specifically in the document granting the rights, ideally by the prior transfer of money to him or by being backed by a guarantee or bond.

16.35 The existence of these rights means that those seeking to invest in land should have regard not only to the condition of the land itself, but also to the condition of adjacent or adjoining land. It would be unfortunate if, having acquired a clean site, the new owner then finds that he has to let it be disrupted to remediate the contaminated site next door. It would be even more unfortunate if he had constructed a development which then had to suffer some form of demolition or restructuring. This is something which needs to be borne in mind when commissioning any form of desk study or environmental assessment.

It is notable, and perhaps regrettable, that the identification procedure which precedes the issue of a remediation notice (see **16.14** above) does not provide for notification to the owners of adjoining or adjacent land. They are not entitled to be served with notice that neighbouring land has been identified as contaminated land, notwithstanding that, at the time of identification, the local authority might have been aware of the need to carry out remediation works on their adjoining or adjacent land. The prudent purchaser will, therefore, ensure that his researches include checking the contaminated land register in respect of neighbouring land, however, whilst this will show remediation notices, remediation declarations or remediation statements which may have been served, it will not help where the local authority has merely identified the neighbouring land as contaminated land but is still in negotiation with the appropriate persons as to voluntary action or is still considering regulatory action. This is because the initial identification of a site as contaminated land is not registered (see **16.33** and **16.35** above) and registration takes place only when a remediation notice, remediation declaration or remediation statement is issued. Those involved in land neighbouring contaminated land are thus disadvantaged by the statutory procedures.

'Appropriate persons'

16.36 A remediation notice must be served on an 'appropriate person' only: s 78E(1). In the first instance, an appropriate person is any person or any of the persons who caused or knowingly permitted any of the contaminative substances to be in, on or under the land: s 78F(2). The phrases 'caused' and 'knowingly permitted' are not new and can be found in, among other legislation, the Water Resources Act 1991 (see **7.7**, **7.10** and **7.16**) and the EPA (see **8.14**). The phrases are discussed at **6.6–6.16**.

If that person cannot be found, after reasonable inquiry, then the current owner or occupier of the contaminated land is an appropriate person in respect of these substances: s 78F(4) and (5). On the basis of existing law in relation to statutory nuisance, the enforcing body would have to make reasonable inquiries to identify the person responsible, but they do not have to make considerable efforts: *Rhymney Iron Co v Gelligaer District Council* [1917] 1 KB 589. Tracing the person responsible under s 78F(2) and (3) involves consideration of two separate concepts: first, actually identifying the source of the contamination (it might be that an owner or occupier becomes liable under s 78(4) and (5) because of the origin of the contamination is unknown); secondly, physically finding the person responsible, having established who he is (he may have left the country and cannot be traced). It is not clear whether the provisions cover both different cases of not being found. Nor is it clear how the provisions operate where the polluter is identified as a company which has since been dissolved or struck off. There the polluter has been identified and traced, but no longer exists.

16.37 The authority may serve different remediation notices on different appropriate persons in consequence of the presence of different substances, in, on or under any land or waters: s 78E(2). Where two or more persons are appropriate persons, then any remediation notices served on them must state the proportion of the costs of remediation which each must bear: s 78E(3). If there is more than one potential appropriate person, then the enforcing authority will have to act in accordance with mandatory guidance from the Secretary of State when determining which shall be treated as appropriate persons and when apportioning costs between them: s 78F(6) and (7).

Liability

16.38 The liability of the appropriate person is, in the first instance, substance specific in the sense that he liable for remediation only in the respect of the contaminative substances he either caused, knowingly permitted or assumed responsibility for as owner or occupier under s 78F(4): s 78F(3) and (5). Generally, he is not liable for remediation in respect of other contaminants at the site – they will be the responsibility of other appropriate persons. There are two qualifications to this principle. First, the liability of a person who caused or knowingly permitted a contaminative substance to be in, on or under any land extends to substances arising from chemical reactions or biological processes affecting the original contaminative substances: s 78F(9). It is not clear whether the same liability extends to owners or occupiers who are generally liable under s 78F(4), because the wording of s 78F(9) is expressly limited to persons who caused or knowingly permitted. Secondly, s 78F(10) provides that things to be done by way of remediation may be regarded as referable to the presence of any substance notwithstanding that they would not have to be done in consequence only of:

(a) the presence of that substance in any quantity; or
(b) the quantity of that substance which any particular person caused or knowingly permitted to be present.

Naturally occurring substances

16.39 A person may become an 'appropriate person' in respect of naturally occurring substances. The statutory definition of 'substance' in s 78A(9) includes natural substances (see **16.6** above) and, as noted above, s 78F(9) includes liability for substances arising from chemical reactions or biological processes. Thus there would be liability for natural substances which are contaminants per se or the contaminative products of natural substances, eg radon gas or methane produced by decomposing peat. Clearly, no person could be an appropriate person under s 78F(2) or (3) on the basis of 'causing' natural contamination unless he actually imports a natural contaminant onto the site. It could be argued that an owner or occupier would have 'knowingly permitted' the creation of substances which arose from a natural chemical reaction or biological process if he had the power to prevent or inhibit the creation of those substances or, under s 78F(4), on the basis that the s 78F(2) person cannot be found.

Controlled waters

16.40 Section 78J provides an important restriction in respect of controlled waters on the liability of owners or occupiers who are, prima facie, liable under s 78F(4) or (5), ie where the person who caused or knowingly permitted the contamination cannot be found. If the land is contaminated land by virtue of s 78A(2)(b) (ie in respect of effect on controlled waters), then (whether or not the provision relating to other forms of contamination in s 78A(2)(a) also applies) that owner or occupier cannot be required to carry out remediation to that or any other land or waters other than that remediation justified solely under s 78A(2)(a): s 78J(2). In other words, Part IIA is read, in this case, as though all references to pollution of controlled waters have been omitted from it.

A similar restriction applies to any person who has permitted or might permit water from a mine abandoned after 31 December 1999 to enter controlled waters, or reach a place which is or was likely (in the opinion of the enforcing authority) to enter such waters. In such a case, Part IIA is again read as though all references to pollution of controlled waters have been omitted from it, and the service of, and terms of, any remediation notice are similarly restricted: see s 78J(3)–(6).

Migration of contaminants

16.41 If a person appears to have caused or knowingly permitted the presence of contaminative substances in, on or under any land, then he will also be liable in respect of any land to which they appear to have escaped (the receiving site being described in Part IIA as 'land A'): s 78K(1). A remediation notice can be served on the owner or occupier of land A in respect of land A only, and can include other land only if he caused or knowingly permitted contamination of that other land: s 78K(3). Under s 78K(4), if the contaminative substances escape from land A onto any further land (called 'land B') then the owner of land A cannot be served with a remediation notice unless:

- he appears to the enforcing authority to have caused or knowingly permitted the contaminative substances to escape to land A; or,
- he is also the owner or occupier of land B.

Section 78K(5) provides that the new owner or occupier of a contaminated site will not be served with a remediation notice in respect of the escape of contaminant substances to other land if the escape was due to an earlier owner or occupier and he (the new owner or occupier) did not cause or knowingly permit the escape. Notwithstanding these provisions, the enforcing authority may enter land affected by the escaped contaminants and carry out remedial measures; however, its ability to recover the costs of doing so are severely limited: s 78K(6).

Appeals against remediation notices

16.42 The recipient of a remediation notice has, under s 78L(1), 21 days to appeal against the notice:

- if it was served by the local authority, to the magistrates' court; or
- if it was served by the Environment Agency, to the Secretary of State.

Regulations may be made to prescribe (among other things) the grounds of appeal, the appeal procedure and cases where remediation notices may be suspended until the appeal is decided: s 78G(4) and (5).

16.43 On appeal, the appellate authority:

(a) shall quash the notice if satisfied that there is a material defect in it; or
(b) may confirm the notice, with or without modification, or quash it: s 78G(2).

The requirement to quash in (a) is mandatory, whereas the power to quash in (b) is discretionary and allows the appellate authority to quash where there is no technical defect in the notice but where it is satisfied that the notice is not justified on the merits.

The appellate authority does not have an independent power to issue fresh remediation notices or to add fresh appropriate persons to existing notices. Beyond that, the appellate authority would appear to stand in the shoes of the enforcing body when considering an appeal, and, as such, would substitute its own discretion for that of the enforcing body instead of quashing or modifying the notice only if convinced that the enforcing authority was in error. That is to say, the appellate authority will exercise a 'de novo' jurisdiction rather than act as a court of review. This approach means that, among other things, the appellate body will be entitled to consider evidence (including hearsay evidence) which was not before the enforcing authority when it made the relevant decisions: see *Kavanagh v Chief Constable of Devon and Cornwall* [1973] 3 All ER 657.

The Secretary of State may, by way of regulations, provide a right of further appeal from a decision of the magistrates' court or tribunal: s 78L(4)(b).

16.44 Appeals cannot sensibly be conducted on the basis that all relevant parties will produce their evidence on the day of the appeal only (in examination-in-chief), because the analysis of technical reports can be done properly only if such reports are disclosed some time before the hearing and proper time is given for their analysis. The usual procedure in tribunals such as town and country planning public inquiries and compulsory purchase public inquiries is that proofs of evidence are exchanged by the parties before the start of the inquiry (normally 3 weeks before) and it would seem sensible that the regulations which prescribe the appeals procedures should include similar requirements. Presumably, the members of the appellate body will be entitled to visit the relevant site.

16.45 The availability of this statutory appeal procedure constrains the use of judicial review as a remedy. First, the statutory appeal process must be exhausted before judicial review can be initiated: see *R v Epping and Harlow General Commissioners ex parte Goldstraw* [1983] 3 All ER 257 at p 262. Secondly, judicial review cannot be used as a way of relitigating the merits of an appeal in any event: see *The Supreme Court Practice 1995* (Sweet & Maxwell, 1994) vol 1, para 53/1–14/13.

16.46 The provision of first-instance appeals to magistrates' courts in non-special cases is puzzling given the considerable legal and technical complexities which such appeals will involve. It is difficult to see how a bench of lay persons, however conscientious, can be expected to reach reasoned decisions on such esoteric subject-matter.

Failure to comply with a remediation notice

16.47 Failure to comply with a remediation notice without reasonable excuse is a criminal offence: s 78M(1). If the contaminated land is industrial, trade or business premises, then the maximum fine is £20,000; otherwise the maximum fine is £5,000: s 78M(3) and (4). There are daily penalties for ongoing offences: s 78M(3) and (4). The phrase 'industrial, trade or business premises' is defined by s 78M(6) in terms identical to s 79(7) of the EPA and the same problems of interpretation will apply: see **9.29**.

The 'reasonable excuse' defence in s 78M(1) follows earlier practice and will be subject to the same constraints and case-law, including that a matter which should have been raised under a statutory appeal cannot be raised as a 'reasonable excuse': see **9.24–9.25**.

16.48 Part IIA does not provide a 'best practicable means' defence (see **9.26**) and, since lack of funds does not amount to a reasonable excuse (see **9.25**), the impecunious defendant is left to raise financial matters by way of mitigation only. The only financial defence appears to be in s 78M(2) which states:

> 'Where the remediation notice in question is one which was required by section 78E(3) above to state, in relation to the requirement which has not been complied with, the proportion of the cost involved which the person charged with the offence is liable to bear, *it shall be a defence for* that person to prove that the only reason why he has not complied with the requirement is that one or more of the other persons who are liable to bear a proportion of that cost refused, or was not able, to comply with the requirement.' (Emphasis added.)

16.49 If the enforcing authority is of the opinion that criminal proceedings would be ineffectual, then it may apply for an injunction in the High Court for the purposes of securing compliance with the remediation notice: s 78E(5).

Remediation by the enforcing authority

16.50 Section 78N provides that the enforcing authority may enter the contaminated land, any controlled waters affected by that land or any land adjoining or adjacent to that land or water for the purposes of remediation:

(a) where the enforcing authority considers it necessary to do anything itself by way of remediation for the purpose of preventing the occurrence of any serious harm, or serious pollution of contaminated waters, of which there is imminent danger (s 78N(3)(a));
(b) where an appropriate person has entered into a written agreement with the enforcing authority for that authority to do, at the cost of that person, that which he would otherwise be required to do under Part IIA by way of remediation (s 78N(3)(b));
(c) where a person on whom the enforcing authority serves a remediation notice fails to comply with any of the requirements of the notice (s 78N(3)(c));
(d) where the enforcing authority is precluded by ss 78J (see **16.40**) or 78K (see **16.41**) from including something by way of remediation in a remediation notice (s 78N(3)(d));
(e) where the enforcing authority considers that, were it to do some particular thing by way of remediation, it would decide, by virtue of s 78P(2) (see **16.52**) or any guidance issued under that subsection:
    (i) not to seek to recover under s 78P(2) any of the reasonable costs incurred by it in doing that thing; or
    (ii) to seek so to recover only a portion of that cost (s 78N(3)(e)); and
(f) where no person has, after reasonably inquiry, been found who is an appropriate person in relation to any particular thing (s 78N(3)(f)).

It is surprising that s 78N(3)(b) is confined to agreements with the appropriate persons only. There may be cases where persons other than appropriate persons are interested in the land and prepared to enter into such an agreement, for example a mortgagee bank which is out of possession.

16.51 Section 78N(4) provides that, subject to s 78E(4) and (5) (see **16.28** above), the things which are appropriate for the enforcing authority to do by way of remediation are:

(a) in a case falling within s 78N(3)(a), anything by way of remediation which the enforcing authority considers necessary for the purpose mentioned in that paragraph;
(b) in a case falling within s 78N(3)(b), anything specified in or determined under the agreement mentioned in that paragraph;
(c) in a case falling within 78N(3)(c), anything which the person mentioned in that paragraph was required to do by virtue of the remediation notice;
(d) in a case falling within s 78N(3)(d), anything by way of remediation which the enforcing authority is precluded by ss 78J or 78K from including in a remediation notice; and

(e) in a case falling within s 78N(3)(e) or (f), the particular thing mentioned in the paragraph in question.

16.52 Section 78P(1) allows the enforcing authority to recover the costs of remediation in certain cases:

> 'Where, by virtue of section 78N(3)(a), (c), (e) or (f) above, the enforcing authority does any particular thing by way of remediation, it shall be entitled, subject to sections 78J(7) [see below] and 78K(6) [see below] ..., to recover the reasonable cost incurred in doing it from the appropriate person or, if there are two or more appropriate persons in relation to the thing in question, from those persons in proportions determined pursuant to section 78F(7) above.'

As to s 78J(7), it will be recalled, from **16.40**, that there is a restriction on the service of remediation notices on owners or occupiers of contaminated land relating to the pollution of controlled waters. Notwithstanding that restriction, s 78J(7) allows the enforcing authority to step in and carry out the remediation, but without the ability to recover the costs of doing so from the owner or occupier. Section 78K(6) has the same effect in respect of the restriction relating to the escape of contaminative substances to neighbouring land or waters: See **16.41**.

Section 78P(2) stipulates that, in deciding whether to recover the cost of remediation, and if so, how much of the cost, the enforcing authority shall have regard:

(a) to any hardship which the recovery may cause to the person from whom the cost is recoverable; and
(b) to any advisory guidance issued by the Secretary of State for this purpose.

Part IIA provides that interest may be added to the costs in certain cases and allows for enforcement of costs and interest by the attachment of a charging notice on the land: s 78P(3) to (13).

ACCESS TO INFORMATION

16.53 Part IIA contains a number of provisions requiring the production of publicly accessible information. Section 78R(1) requires every enforcing authority to maintain a register including prescribed particulars of or relating to:

- remediation statements or remediation;
- remediation notices served by that authority;
- appeals against such notices;
- appeals against charging notices;
- notices designating, or terminating the designation of, special sites;
- notices by the recipients of remediation notices, person who issued remediation statements or from the owners or occupiers of land subject to a remediation notice or remediation statement of what they claim has been done by way of remediation;

- convictions under Part IIA; and
- such other matters as may be prescribed.

The Environment Agency must, also copy those particulars to the local authority in whose area the contaminated land is situated: s 78R(4).

16.54 The enforcing authority is not required to note the identification of land as contaminated land prior to the issue of a remediation notice, remediation declaration or remediation statement. This is a deliberate omission (see **16.15** above), but it puts potential purchasers of that land or neighbouring land at a disadvantage. One approach might be by way of an application to the enforcing authority under the Environmental Information Regulations 1992 (SI 1992/3240) specifically asking whether or not an identification notice has been served on a specific site: see Chapter 15. Another approach is that the potential purchaser must be careful to address the issue in his inquiries of the vendor before the purchase: see Chapter 14 and Model 1 in Appendix 1.

Also, the Environment Agency is required to publish a report on the state of contaminated land in England and Wales from time to time or if the Secretary of State at any time so requests: s 78U(1). If the Environment Agency so requests in writing, a local authority must furnish it with such information as the authority may have or may reasonably be expected to obtain in respect of the exercise of its functions in relation to Part IIA of the EPA: s 78U(2).

RECEIVERS AND INSOLVENCY PRACTITIONERS

16.55 There is a sting in the tail of Part IIA for any receiver or insolvency practitioner who is deemed to be a 'person acting in a relevant capacity'. The phrase 'person acting in a relevant capacity' is defined by s 78X(4) as meaning:

> '(a) a person acting as an insolvency practitioner, within the meaning of s 388 of the Insolvency Act 1986 (including that section as it applies in relation to an insolvent partnership by virtue of any order made under s 421 of that Act);
> (b) the Official Receiver acting in a capacity in which he would be regarded as acting as an insolvency practitioner within the meaning of s 388 of the Insolvency Act 1986 if subs (5) of that section were disregarded;
> (c) The Official Receiver acting as receiver or manager;
> (d) a person acting as a special manager under ss 177 or 370 of the Insolvency Act 1986;
> (e) ...
> (f) a person acting as a receiver or receiver and manager;
> (i) under or by virtue of any enactment; or,
> (ii) by virtue of his appointment as such by an order of a court or by any other instrument.'

Item (f)(i) will include receivers appointed under the Law of Property Act 1925.

In respect of civil liability for the costs of remediation, s 78X(3)(a) provides that a person acting in a relevant capacity:

> '... shall not thereby be personally liable, under [Part IIA], to bear the whole or any part of the cost of doing any thing by way of remediation, unless that thing is to any extent referable to substances whose presence in, on or under the contaminated land in question is a result of any act done or omission made by him which it was unreasonable for a person acting in that capacity to do or make.'

In respect of criminal liability, s 78X(3)(b) provides that a person acting in a relevant capacity:

> '... shall not thereby be guilty of an offence under or by virtue of s 78M above unless the requirement which has not been complied with is a requirement to do some particular thing for which he is personally liable to bear the whole or any part of the cost.'

Section 78X(5) adds that regulations may make different provision for different cases or circumstances.

ABANDONED MINES

16.56 Sections 58 and 60 of the Environment Act provide amendments to the law in relation to abandoned mines. As discussed at **2.16**, many abandoned mines are an ongoing source of pollution, yet, historically, the owners of abandoned mines have been excluded from statutory liability in respect of pollution of controlled waters. Section 85 of the Water Resources Act 1992 (WRA) provides criminal offences connected with the pollution of controlled waters, but discharges from abandoned mines fall outside its ambit (see **7.12**). Section 161 of the WRA allows the National Rivers Authority to remedy pollution caused by abandoned mines, but not to recover the costs from the operators: **7.15–7.16**.

Section 58 of the Environment Act inserts a new Chapter IIA into the WRA. Section 91B(1) of the WRA now provides that the operator of a mine must give notice of a proposed abandonment to the Environment Agency at least 6 months before the abandonment takes effect. Failure to give such notice is a criminal offence unless it takes place in an emergency in order to avoid danger to life or health and notice is then given as soon as reasonably practicable after the abandonment has happened: ss 92B(3) and (4). Regulations may prescribe the information to be contained in an abandonment notice: s 91B(2). Section 91B(7) provides that where the Environment Agency receives an abandonment notice or otherwise learns of an abandonment or proposed abandonment and considers that, in consequence of the abandonment or proposed abandonment, any land is likely to become contaminated land, then the Agency must inform the relevant local authority.

Section 60 of the Environment Act amends ss 89 and 161 of the WRA (see **7.12** and **7.15–7.16**). The defence against a s 85 prosecution for operators of abandoned mines, in s 89(3), will no longer be available in respect of mines abandoned after 31 December 1999: s 89(3A). Section 161 of the WRA is amended so that the operators of mines abandoned after 31 December 1999 will become liable for expenses incurred by the Environment Agency in remedial works or operations in respect of pollution of controlled waters from such mines.

WORKS NOTICES UNDER THE WATER RESOURCES ACT 1991

16.57 Schedule 22 of the Environment Act amends s 161 of the WRA (see **7.15–7.18**) by the introduction of a 'works notice' procedure. Section 161(1) of the WRA is amended by a new s 161(1A) (inserted by para 161(3) of Sch 22) which provides that:

> 'Without prejudice to the power of the Environment Agency to carry out investigations under section 161(1) above, the power conferred by that subsection to carry out works and operations shall only be exercisable in a case where:
>
> (a) the Agency considers it necessary to carry out forthwith any works or operations falling within paragraph (a) or (b) of that subsection; or
> (b) it appears to the Agency, after reasonable inquiry, that no person can be found on whom to serve a works notice under section 161A below.'

Paragraph 162 of Sch 22 inserts a new s 161A(1) into the WRA which provides that, subject to s 161A, where it appears to the Agency that any poisonous, noxious or polluting matter or any solid waste matter is likely to enter, or to be or to have been present in, any controlled waters, the Agency shall be entitled to serve a works notice on any person who, as the case may be:

(a) caused or knowingly permitted the matter in question to be present at the place from which it is likely, in the opinion of the Agency, to enter any controlled waters; or
(b) caused or knowingly permitted the matter in question to be present in any controlled waters.

The amendments made to the Environment Act provide rights of appeal against works notice: s 161C. Failure to comply with a works notice is a criminal offence: s 161D.

16.58 A new s 161A(2) defines a 'works notice' as a notice requiring the person on whom it is served to carry out such of the following works or operations as may be specified in the notice, that is to say:

(a) in a case where the matter in question appears likely to enter any controlled waters, works or operations for the purpose of preventing it from doing so; or,
(b) in a case where the matter appears to be or to have been present in any controlled waters, works or operations for the purpose;

(i) of removing or disposing of the matter;
(ii) of remedying or mitigating any pollution caused by its presence in the waters; or
(iii) so far as it is reasonably practicable to do so, of restoring the waters, including any flora and fauna dependent on the aquatic environment of the waters, to their state immediately before the matter became present in the waters.

A works notice must specify the periods within which the person on whom it is served is required to do each of the things specified in the notice: s 161A(3). The Agency shall, before serving a works notice, reasonably endeavour to consult the proposed recipient: s 161A(4). The service of works notices is restricted in respect of abandoned mines: ss 161A(8), (9) and (10).

Granting of rights

16.59 New s 161B(1) provides that a works notice may require a person to carry out works or operations in relation to any land or waters notwithstanding that he is not entitled to carry out those works or operations. And s 161B(2) adds that any person whose consent is required before any works or operations required by a works notice may be carried out shall grant, or join in granting, such rights in relation to any land or waters as will enable the person on whom the works notice is served to comply with any requirements imposed by the works notice. These rights are identical to those in respect of remediation notices and the same comments apply as to the position of those affected, particularly neighbouring landowners: see **16.33**.

The impact of these rights is ameliorated by s 161B(3) which provides that before serving a works notice, the Agency shall reasonably endeavour to consult every person who appears to it:

(a) to be the owner or occupier of any relevant land; and
(b) to be a person who might be required by s 161B(2) above to grant, or join in granting, any rights, concerning the rights which that person may be so required to grant.

The Environment Act gives the person who grants, or joins in granting, any rights pursuant to s 161B(2) an ability to claim compensation from the person on whom the works notice was served (s 161B(5)) and this power is subject to the same criticisms which apply to the analogous remediation notice provision: see **16.34**).

WASTE REGULATION

16.60 Schedule 22 of the Environment Act makes a large number of detailed amendments to Part III of the EPA in respect of waste regulation: see generally Chapter 8. Importantly the definition of 'waste' in s 75(2) of the EPA (see **8.4**) is amended by para 88(2) of Sch 22 so that 'waste' means:

> 'any substance or object in the categories set out in Schedule 2B [see below] to this Act which the holder discards or intends or is required to discard; and for the purposes of this definition:
>
> > "holder" means the producer of the waste or the person who is in possession of it; and
> > "producer" means any person whose activities produce waste or any person who carries out pre-processing, mixing or other operations resulting in a change in the nature or composition of this waste.'

16.61 The presumption that anything discarded is waste unless the contrary is proved in s 75(3) (see **8.4**) will cease to have effect: para 88(3).

In order to bring the UK definition of waste into line with the Waste Directive definition, a new Sch 2B is added to the EPA which lists the categories of waste as follows:

(a) production or consumption residues not otherwise specified below;
(b) off-specification product;
(c) products whose date for appropriate use has expired;
(d) materials spilled, lost or having undergone other mishap, including any materials, equipment, etc, contaminated as a result of the mishap;
(e) materials contaminated or soiled as a result of planned actions (eg residues from cleaning operations, packing materials, containers, etc);
(f) unusable parts (eg reject batteries, exhausted catalysts, etc);
(g) substances which no longer perform satisfactorily (eg contaminated acids, contaminated solvents, exhausted tempering salts, etc);
(h) residues of industrial process (eg slags, still bottoms, etc);
(i) residues from pollution abatement processes (eg scrubber sludges, baghouse dusts, spent filters, etc);
(j) machining or finishing residues (eg lathe turnings, mill scales, etc);
(k) residues from raw materials extraction and processing (eg mining residues, oil field slops, etc);
(l) adulterated materials (eg oils contaminated with PCBs, etc);
(m) any materials, substances or products whose use has been banned by law;
(n) products for which the holder has no further use (eg agricultural, household, office, commercial and shop discards, etc);
(o) contaminated materials, substances or products resulting from remedial action with respect to land;
(p) any materials, substances or products which are not contained in the above categories.

Appendix 1
MODEL DOCUMENTS

1 PRELIMINARY INQUIRIES

[The material set out below is discussed in chapter 14. It is envisaged that these preliminary inquiries would be accompanied by a glossary of defined terms: such a glossary is provided in Model 5.]

1 Please list and briefly describe the activities which are currently and which have in the past been carried out on or at the property.

2 Have any activities which use or produce or involve:

(a) processes prescribed by s 2 of the Environmental Protection Act 1990;
(b) hazardous substances as specified in any regulation made under the Planning (Hazardous Substances) Act 1990;
(c) 'special waste' as defined in s 62 of the Environmental Protection Act 1990, the Control of Pollution (Special Waste) Regulations 1980 or associated regulations; or
(d) radioactive material or waste as defined in s 18 of the Radioactive Substances Act 1960

been carried out any time on or at the property by the vendor or by any predecessor in title of the vendor or on or at any land within 250 m of the property?

3 Has the property or any building, tank or vessel situate on, in or under it at any time ever been used for the deposit, burial, storage or incineration of any waste chemicals or other substance?

4 If the answer to either **2** or **3** above is yes, please supply full details including the date of the activity, the type and quantity of substances involved and a plan showing the exact location of the activity on the property or other premises and copies of all relevant consents, licences or permissions.

5 Has any land within 250 m of the property ever been used for the deposit, burial, storage, treatment or incineration of any waste chemicals or other substances?

If yes, please supply full details including the date, type and quantity of the waste chemicals or other substances, the method of disposal or treatment and a plan showing the area used.

6 Is the vendor aware of:

(a) any quarrying or mineral extraction operations being carried out; or

(b) any proposal to create a site for the deposit of waste

within 400 m of the property?

If so, please supply full details together with a plan showing the location.

7 Is the property now or has it ever been Contaminated or the source of any Pollution?

If so, please give full details.

8 Is the vendor aware of any land or watercourse within 250 m of the property which is or many be Contaminated or the source of Pollution?

If so, please give full details.

9 Has the presence of any landfill gases or other noxious gases been noted or recorded on, in, over or under the property or on, in, over or under any other land or watercourse within 250 m of the property?

If so, please give full details.

10 Is the vendor aware of any Contamination of the property or the groundwater beneath the property or neighbouring property or watercourses passing through under or close to the property?

If so, please give full details.

11 Are there or have there at any time been any underground storage tanks at the property or any property within 250 m of the property?

If yes, please indicate on a plan their precise locations, what they are or were used for and their age.

If they are disused, please give details of the filling materials used or any other method of sealing.

12 Has any part of the property or any nearby property or watercourses possessed, owned or controlled by the vendor ever been the subject of any sampling, testing, site investigation, assessment or audit carried out to test that property's soil, surface water, groundwater or atmosphere or any discharges or emissions from it?

If so, please supply copies of any such survey, environmental audit or assessment or other report in respect of the property.

13 Has any part of the property been subject to work designed to remove or treat contamination or prevent or reduce Pollution?

If yes, please supply details of the works carried out, the contractor who undertook the work and any indemnities, warranties and guarantees relating to that work.

14 Have any excavations on any part of the property been infilled or has the ground on any part of the property been raised?

If yes, please supply details of the location and composition of materials used.

15 Is the property affected by radon gas or in any area designated as a radon-affected area?

If so, please give full details.

16 Has any water, effluent, substance or discharge from the property ever entered or is entering any watercourse, drain or sewer?

If so, please give full details including supplying copies of all relevant correspondence, consents, permissions or other authorisations.

17 Is there on the property any substance (whether inert or otherwise) which may at any time in the future enter any watercourse?

If so, please give full details.

18 Has the vendor or any predecessor in title at any time produced, handled, treated, kept, deposited or disposed of waste on the property so that a licence or other consent for such activities is or was required?

If so, please give full details including copies of any correspondence and/or licences held.

19 What types and quantities of waste are typically produced at the property?

20 What arrangements are in place for the collection of waste from the property?

21 Please confirm that the vendor has not received any complaints about the production, treatment, handling, storage, deposit or disposal of waste on, in or from the property.

22 Has the National Rivers Authority or the Environment Agency exercised any powers specified in s 161 of the Water Resources Act 1991 or its predecessor Acts in relation to the property and/or is the vendor aware of any circumstances which might lead to exercise of such powers?

If so, please give full details.

23 Has any Regulatory Body or any other statutory body:

(a) ever served or made any notice or order in relation to the property, or ever intimated that such notice or other might be served or made;
(b) ever taken any action against any person in relation to the property including any owner or occupier of the property or any person acting for or employed by any such owner or occupier or ever intimated that such action might be taken;
(c) ever exercised any power of entry in relation to the property?

24 Please confirm that neither the vendor nor any predecessor in title has received any complaint and that there have been no disputes about the use of the property, noise generated on the property, the release of any substances from the property or other matters defined as statutory nuisances in s 79 of the Environmental Protection Act 1990.

25 Are there any claims, investigations or other proceedings against or threatened against the vendor in relation to breaches of any Environmental Law or Environmental Code or damage caused as a result of a condition of the property or activities carried out on the property?

If so, please give full details.

26 Please confirm that the purchaser may take any samples and/or carry out any tests which the purchaser requires on the property and have such samples analysed.

Please give the names, addresses and telephone numbers of the person or persons to contact to arrange an appointment for this.

2 ENVIRONMENTAL WARRANTIES

[The material set out below is discussed in Chapter 14. It is envisaged that these environmental warranties would be accompanied by a glossary of defined terms: such a glossary is provided in Model 5.]

1 The vendor has at all times complied with all Environmental Laws and Environmental Codes applicable to or applied to the property and has obtained, and complied with, all permissions, authorisations and consents required by or pursuant to such Environmental Laws and Environmental Codes in relation to the property.

2 So far as the vendor is aware, all previous owners and occupiers of the property have complied with all Environmental Laws and Environmental Codes applicable to or applied to the property and have obtained and complied with all permissions, authorisations and consents required by or pursuant to such Environmental Laws and Environmental Codes in relation to the property.

3 Neither the property nor any land within 200 m of the property is now or has ever been Contaminated or the source of Pollution.

3 ENVIRONMENTAL INDEMNITY

[The material set out below is discussed in Chapter 14. It is envisaged that this environmental indemnity would be accompanied by a glossary of defined terms: such a glossary is provided in Model 5.]

1 The vendor shall indemnify and keep indemnified [until the site investigation has been carried out and the report arising from the site investigation received by the purchaser] [until the planning permission has been granted] the purchaser against all [reasonable] losses, costs, damages, expenses and liabilities incurred by the purchaser [from the completion date] in respect of any claim by any person or Regulatory Body arising from contamination or pollution of or arising from the property including, but not limited to, the costs and fees incurred in remedial or restoration works or operations or the removal from or securing on the property of any substance, liquid, gas or vapour.

2 The vendor shall have no liability under Clause 1 to the extent that the claim is due to the act or omission of the purchaser or his agent or employee [or due to any retrospective change in any Environmental Law or Environmental Code] [or due any event occurring after the completion date].

3 The vendor's liability under Clause 1 shall be limited to the purchase price whether such liability occurs by a single claim or the aggregate of separate claims under Clause 1.

4 [Claims procedure: see Model 4]

4 CLAIMS PROCEDURE

[The material set out below is discussed in Chapter 14. It is envisaged that this claims procedure would be accompanied by a glossary of defined terms: such a glossary is provided in Model 5.]

1 The vendor shall have no liability in respect of clauses [ ] above unless the purchaser duly observes and complies with the claims procedure set out in clauses [ ] below.

2 The purchaser shall notify the vendor of the receipt of any notice, order or letter from any person or Regulatory Body relating to the breach of or any Environmental Law or Environmental Code or making any claim in respect of or request for payment under any Environmental Law or Environmental Code or intimating that such claim or request will or may be made as soon as is reasonably practicable after receipt of the same and in any event within 7 days thereof.

3 In the event of a claim by the purchaser pursuant to clauses [ ] [ie claims from third parties or regulatory bodies]:

(a) the purchaser shall supply the vendor with all correspondence, reports and other documents relevant to and arising in respect of the claim and supply the vendor with full particulars thereof;
(b) the vendor shall be entitled to negotiate directly with any person or regulatory body making, giving rise to or relevant to the claim and to take over and conduct the defence or settlement of the claim and shall have full discretion in the conduct of any proceedings or in the settlement of the claim and the purchaser shall give the vendor all such information and assistance as the vendor may reasonably require; and
(c) no admission, offer, promise or indemnity shall be made or given by or on behalf of the purchaser in respect of any claim without the consent of the vendor.

4 In the event of any claim by the purchaser pursuant to clauses [ ] above [ie a claim by the purchaser against the vendor only]:

(a) the vendor and purchaser shall agree the steps to be taken to investigate and remedy the matters giving rise to the claim;
(b) the vendor and purchaser shall agree the indemnity and terms of engagement of any contractor commissioned to carry out any investigations, operations or works in respect of the claim and agree any instructions which may from time to time be given to that contractor; and
(c) if any dispute arises between the vendor and purchaser as to the

basis for the claim or the matters mentioned at (a) and (b) above, then the dispute shall be referred for final determination at the request of either party to an independent chartered surveyor, having experience appropriate to the conditions giving rise to the claim, agreed between the parties, or in default of such agreement nominated by the President for the time being of the Institute of Chartered Surveyors upon the request of either party, and such independent chartered surveyor shall act as an expert and not as an arbitrator and his fees shall be borne as he shall direct.

5 GLOSSARY OF DEFINED TERMS

[The material set out below is discussed in Chapter 14. It is envisaged that this glossary would accompany Models 1–4.]

For the purpose of [this warranty] [this indemnity] [these preliminary enquiries] the following phrases shall where the context admits have the following meanings:

'Contamination' means the presence on in or under any land or on or in water or in the air of any substance, liquid or gas not naturally occurring there.

'Environmental Code' means any code of practice, circular, planning policy guidance note, waste-management paper, or advice issued by any department or ministry of Her Majesty's Government relating to the prevention or regulation of pollution or contamination.

'Environmental Law' means any law regulation or by-law relating to the prevention or regulation of pollution or contamination arising from or prescribed by any judgment of the Supreme Court of Judicature or any directive or regulation produced by the [European Community] or any Act of Parliament including (but not limited to) the Public Health Act 1936, the Control of Pollution Act 1974, the Environmental Protection Act 1990, the Town and Country Planning Act 1990 and the Water Resources Act 1991.

'Pollution' means the release or escape to land, water or air of any substance, liquid or gas not naturally occurring there which is capable of causing harm to the health of any living organism or other interference with the ecological systems of which they form part and in the case of man includes offence to any of his senses or harm to his property.

'Regulatory body' means any local authority or waste-regulation authority, the National Rivers Authority, Her Majesty's Inspectorate of Pollution or any other statutory body possessing legal powers in relation to pollution or contamination.

'Watercourse' includes all rivers, streams, ditches, drains, cuts, culverts, dykes or sluices and sewers and passages through which water flows, except mains and other pipes which are used by a water undertaker for the purpose only of providing a supply of water to any premises.

Appendix 2

MATERIALS

1 WATER COURSES AFFECTED BY DISCHARGES FROM ABANDONED COAL MINES AND ASSOCIATED SPOIL TIPS

[The material set out below, taken from *Abandoned Mines and the Water Environment* (NRA, Water Quality Series No 14), is discussed in Chapter 2.]

NORTHUMBRIAN REGION

(Tipalt Burn)
(River South Tyne)
Elsdon Burn (River North Tyne)
River Gaunless (River Wear)
South Burn (River Wear)
Old Durham Beck (River Wear)

NORTH WEST REGION

Trent and Mersey Canal
Bridgewater Canal
Pendle Water
River Brun
Green Brook
Black Clough
River Calder
Copy Clough
Everage Clough
White Ash Brook
Hyndbun Brook
Lottuce Brook
Woodnook Water
Valley Brook
River Dane
River Irwell
River Sett
Lydgate Brook
Ealees Brook
Naden Brook
Small Brook
Cutacre Brook
Hindsford Brook
Ghyll Beck
Westnewton Beck
River Keekle

SEVERN TRENT REGION

Cannop Brook (tributary of the River Severn)
River Amber (tributary of the River Derwent)
River Wye (tributary of the River Derwent)

SOUTH WEST REGION

Restronguet Creek
Carrick Roads
Falmouth Bay

WELSH REGION

River Tawe
River Teifi
River Clyne
River Rhymney
River Sirhowy
River Ebbw
River Pelenna

YORKSHIRE REGION

River Don
River Little Don
River Drone
River Rother
River Doe Lea
River Dearn
River Hipper
River Sheaf
River Calder
River Holme
River Cononley Beck

Region	Discharges by number	km affected	Working	Abandoned
Northumbria	20	43	2	17
North West	5	36	1	4
Severn Trent	2	5		All
South West	Unknown	212	1	About 1700
Welsh	44	114	1	43
Total	>71	410		

Discharges from metal mines which cause significant pollution problems (by original NRA region)

2 WASTE MANAGEMENT LICENSING REGULATIONS 1994, SCHEDULE 4

[The material set out below is discussed in Chapter 8.]

Part II

Substances or objects which are waste when discarded etc.

1 Production or consumption residues not otherwise specified in this Part of this Schedule (Q1).

2 Off-specification products (Q2).

3 Products whose date for appropriate use has expired (Q3).

4 Materials spilled, lost or having undergone other mishap, including any materials, equipment, etc, contaminated as a result of the mishap (Q4).

5 Materials contaminated or soiled as a result of planned actions (eg residues from cleaning operations, packing materials, containers, etc) (Q5).

6 Unusable parts (eg reject batteries, exhausted catalysts, etc) (Q6).

7 Substances which no longer perform satisfactorily (eg contaminated acids, contaminated solvents, exhausted tempering salts, etc) (Q7).

8 Residues of industrial processes (eg slags, still bottoms, etc) (Q8).

9 Residues from pollution abatement processes (eg scrubber sludges, baghouse dusts, spent filters, etc) (Q9).

10 Machining or finishing residues (eg lathe turning, mill scales, etc) (Q10).

11 Residues from raw materials extraction and processing (eg mining residues, oil field slops, etc) (Q11).

12 Adulterated materials (eg oils contaminated with PCBs, etc) (Q12).

13 Any materials, substances or products whose use has been banned by law (Q13).

14 Products for which the holder has no further use (eg agricultural, household, office, commercial and shop discards, etc) (Q14).

15 Contaminated materials, substances or products resulting from remedial action with respect to land (Q15).

16 Any materials, substances or products which are not contained in the above categories (Q16).

(Note:– the reference in brackets at the end of each paragraph of this Part of this Schedule is the number of the corresponding paragraph in Annex I to the Directive [75/442/EEC].)

PART III

Waste disposal operations

1 Tipping of waste above or underground (eg landfill, etc) (D1).

2 Land treatment of waste (eg biodegradation of liquid or sludge discards in soils, etc) (D2).

3 Deep injection of waste (eg injection of pumpable discards into wells, salt domes or naturally occurring repositories, etc) (D3).

4 Surface impoundment of waste (eg placement of liquid or sludge discards into pits, ponds or lagoons, etc) (D4).

5 Specially engineered landfill of waste (eg placement of waste into lined discrete cells which are capped and isolated from one another and the environment, etc) (D5).

6 Release of solid waste into a water body except seas or oceans (D6).

7 Release of waste into seas or oceans including seabed insertion (D7).

8 Biological treatment of waste not listed elsewhere in this Part of this Schedule which results in final compounds or mixtures which are disposed of by means of any of the operations listed in this Part of this Schedule (D8).

9 Physico-chemical treatment of waste not listed elsewhere in this Part of this Schedule which results in final compounds or mixtures which are disposed of by means of any of the operations listed in this Part of this Schedule (eg evaporation, drying, calcination, etc) (D9).

10 Incineration of waste on land (D10).

11 Incineration of waste at sea (D11).

12 Permanent storage of waste (eg emplacement of containers in a mine, etc) (D12).

13 Blending or mixture of waste prior to the waste being submitted to any of the operations listed in this Part of this Schedule (D13).

14 Repackaging of waste prior to the waste being submitted to any of the operations listed in this Part of this Schedule (D14).

15 Storage of waste pending any of the operations listed in this Part of this Schedule, but excluding temporary storage, pending collection, on the site where the waste is produced (D15).

(Note:– the reference in brackets at the end of each paragraph of this Part of this Schedule is the number of the corresponding paragraph in Annex IIA to the Directive [75/442/EEC].)

PART IV

Waste recovery operations

1 Reclamation or regeneration of solvents (R1).

2 Recycling or reclamation of organic substances which are not used as solvents (R2).

3 Recycling or reclamation of metals and metal compounds (R3).

4 Recycling or reclamation of other inorganic materials (R4).

5 Regeneration of acids or bases (R5).

6 Recovery of components used for pollution abatement (R6).

7 Recovery of components from catalysts (R7).

8 Re-refining, or other reuses, of oil which is waste (R8).

9 Use of waste principally as a fuel or for other means of generating energy (R9).

10 Spreading of waste on land resulting in benefit to agriculture or ecological improvement, including composting and other biological transformation processes, except in the case of waste excluded under Article 2(1)(b)(iii) of the Directive (R10).

11 Use of wastes obtained from any of the operations listed in paragraphs 1 to 10 of this Part of this Schedule (R11).

12 Exchange of wastes for submission to any of the operations listed in paragraphs 1 to 11 of this Part of this Schedule (R12).

13 Storage of waste consisting of materials intended for submission to any operation listed in this Part of this Schedule, but excluding temporary storage, pending collection, on the site where it is produced (R13).

(Note:– the reference in brackets at the end of each paragraph of this Part of this Schedule is the number of the corresponding paragraph in Annex IIB to the Directive [75/442/EEC].)

3 EXTRACTS FROM THE *WDA MANUAL ON THE REMEDIATION OF CONTAMINATED LAND*

[The material set out below is discussed in Chapter 13.]

Box 3.5: Content of a Preliminary Investigation Report

Introduction
- Terms of reference
- Summary of previous work (including any former site investigation or other contamination studies)
- Site description and proposed end use

Information Sources
- Topographical, geological, hydrological data
- Aerial photographs, archive and other records of the site
- Consultation data from third parties

Historical Uses
- Nature and location of previous facilities and services
- Underground mining and workings
- Excavation and landfilling activities
- Decommissioning and decontamination measures

Site Characterisation
- Grid reference and site plan (including boundary definition)
- Topography and site setting
- Hydrological, geological and climate factors
- Built environment and site services
- Regulatory framework

Hazard Identification
- Potential contaminants due to former uses
- Potential for contaminant migration from the site
- Potential contaminative uses adjacent to site
- Potential for contaminant migration into the site

Hazard Assessment
- Potential hazards posed by the site
- Inventory of targets potentially at risk
- Further investigation requirements
- Immediate or interim remedial requirements

Recommendations for Further Work
- Scope
- Phasing
- Precautions (including environmental protection measures)
- Costs
- Health and safety requirements

Appendices
- Supporting data for the above

Box 3.11: Examples of common investigation techniques

Surface Sampling
- Spot samples
- Surface scrapes

Excavations
- Trial pits
- Trial trenches

Borings
- Probes and augers
- Percussion drilling
- Rotary drilling

Geophysical Techniques
- Electrical
- Seismic reflection/refraction
- Gravimetric and magnetic
- Ground penetrating (impulse) radar
- Down-the-hole logging

Vapour and Gas Surveys
- Walkover surveys
- Probes/searcher bars
- Permanent monitoring wells

Groundwater
- Standpipes and piezometers
- Nested and clustered wells

Topographic Surveying
- Technical guidance on these techniques is provided in BS5930 (general ground investigation) and DD 175 (contaminated land investigation).

Box 3.17: Content of a Final Investigation and Assessment Report

Introduction

- Terms of reference
- Summary of previous work (including the Preliminary Investigation Report)
- Site description and proposed end use
- Constraints on the investigation or assessment
- General field conditions at the time of the investigation

Strategy and Methods

- Scope of works
- Programme and any variations
- Methods used
- Health and safety provision
- Environmental protection measures
- Quality assurance plan

Field Work

To include a description of rationale and methods used to locate sampling positions, collect/handle/store samples: conduct in situ testing etc. during:

- Inspection and testing (eg spiking surveys, surface sampling, inventories)
- Ground investigation (eg trial pits, trenches, boreholes (shell & auger, rotary), monitoring installations)
- Supplementary investigation (eg long-term monitoring programmes)

Sampling and Analysis

- Methods and reference standards used
- Quality assurance and control
- COSHH assessment

Chain of Custody Record

- Sample receipt and storage
- Sample preparation and sub-sampling
- Sample analysis
- Sample results
- Sample disposal

Box 3.17: Content of a Main Site Investigation and Assessment

Observations
- General ground conditions (including field description of samples)
- Geological/hydrological regime
- Other features (eg flora/fauna)

Laboratory Reporting
- Analytical results
- Variations to analytical programme
- Accuracy, sensitivity, precision, bias
- Sample retention/disposal

Discussion of Results
- Results of field investigation(s)
- Results of laboratory analyses
- Limitations of the data (if any)

Assessment of Findings
To include a detailed discussion of the rationale, procedures followed and reference data used to conduct:

- Hazard assessment
- Risk estimation
- Risk evaluation

Conclusions and Recommendations
- Degree and extent of contamination on the site
- Significance of observed levels of contamination
- Recommendations on appropriate course of action
- Specification of CROs where remedial action is recommended

Appendices
- Location map and site plan
- Plan of sampling locations
- Graphic logs, boreholes etc. (including installation details)
- Monitoring and analytical data
- Photographic records

4 LIST OF PUBLIC BODIES TO WHICH EC DIRECTIVE 90/313 APPLIES*

[The material set out below is discussed in Chapter 15]

Advisory Committee on Business and the Environment
Advisory Committee on Dangerous Substances
Advisory Committee on Genetic Modification
Advisory Committee on Hazardous Substances
Advisory Committee on Overseas Economic and Social Research
Advisory Committee on the Safety of Nuclear Installations
Advisory Committee on Toxic Substances
Advisory Council on Science and Technology
Advisory Panel on Deregulation
Agriculture Development Advisory Service
Agricultural and Food Research Council
Agriculture Scientific Services
Agricultural Statistics Consultative Committee
Agricultural Training Board
Agricultural Advisory Panel for Wales
Animal Procedures Committee
Assessment Panel for Design and Specification Research
Aviation Committee
Black Country Urban Development Corporation
Black Country Limestone Advisory Panel
Bristol Urban Development Corporation
British Board of Agrement
British Coal Corporation
British Council
British Rail
British Society for Horticultural Research
British Waterways Board
Broads Authority
Building Regulations Advisory Committee
Building Research Establishment
Building Standards Advisory Committee
Cable Authority
Cardiff Bay Development Corporation
Central Manchester Urban Development Corporation
Central Science Laboratory
Central Scotland Water Development Board
Central Statistical Office
Central Transport Consultative Committee
Central Veterinary Laboratory

*Source: DoE Consultation Paper 'The Government's Proposals for the Implementation in United Kingdom Law of the EC Directive on the Freedom of Access to Information on the Environment' (January 1992).

Chemical and Biological Defence Establishment
Civil Aviation Authority
Committee on the Carcinogenicity of Chemicals in Food, Consumer Products and the Environment
Committee on Medical Aspects of Contamination of Air, Soil and Water
Committee on the Medical Aspects of Radiation in the Environment
Committee on the Medical Effects of Air Pollution
Committee on Mutagenicity of Chemicals in Foods, Consumer Products and the Environment
Committee on the Toxicity of Chemicals in Food, Consumer Products and the Environment
Commission for New Towns
Common Council of the City of London
Community Councils (Wales)
Consultative Committee on Freshwater Fisheries in Scotland
Consultative Panel on Badgers and Tuberculosis
Consumer and Vehicle Markets Advisory Committee
Council of the Isles of Scilly
Countryside Commission
Countryside Commission for Scotland (to become Scottish Natural Heritage)
Countryside Council for Wales
County and District Councils in England and Wales
Crown Agents
Crown Estate
Cumbernauld Development Corporation
Customs and Excise
Defence Animal Centre
Department of Agriculture (Northern Ireland)
Department of Education and Science
Department of Energy
Department of the Environment
Department of the Environment (Northern Ireland)
Department of Health
Department of Trade and Industry
Department of Transport
Development Board for Rural Wales
Development Commission
District Councils in Northern Ireland
Domestic Coal Consumers' Council
Drainage Council for Northern Ireland
Drinking Water Inspectorate
East Kilbride Development Corporation
Economic and Social Research Council
English Industrial Estates Corporation
English Nature
English Tourist Board
Enterprise Ulster
Expert Panel on Air Quality Standards

Fire Authority for Northern Ireland
Fisheries Conservancy Board for Northern Ireland
Foreign and Commonwealth Office
Forensic Science Service
Forestry Commission
Foyle Fisheries Commission
Gas Consumers' Council
Glenrothes Development Corporation
Health and Safety Agency for Northern Ireland
Health and Safety Commission
Health and Safety Executive
Highlands and Islands Enterprise Board
Highlands and Islands Fire Board
Hill Farming Advisory Committee for England, Wales and Northern Ireland
Horticultural Research International
Housing Action Trusts
Hydrographic Office
Information Technology Advisory Board
Institute of Development Studies
Internal Drainage Boards
Irvine Development Corporation
Joint Nature Conservation Committee
Laboratory of Government Chemist
Laganside Development Corporation
Lake District Special Planning Board
Land Authority for Wales
Leeds Urban Development Corporation
Letchworth Garden City Corporation
Livingston Development Corporation
London Borough Councils
London Docklands Urban Development Corporation
London Regional Transport
London Residuary Body
London Waste Regulation Authority
Lothian and Borders Fire Board
Macaulay Land Use Research Institute
Medical Research Council
Medicines Commission
Merseyside Urban Development Corporation
Meteorological Committee
Meteorological Office
Meteorological Research Sub-Committee
Metropolitan District Councils
Metropolitan Fire Authorities in England
Milton Keynes Development Corporation
Ministry of Agriculture, Fisheries and Food
Ministry of Defence
National Advisory Group on Environmental Labelling

National Biological Standards Board
National Consumer Consultative Council
National Engineering Laboratory
National Health Service Estates
National Physical Laboratory
National Radiological Protection Board
National Research Development Corporation
National Rivers Authority
Natural Environment Research Council
Natural Resources Institute
Nature Conservancy Council for Scotland (to become Scottish Natural Heritage)
Non-agricultural Pesticides Panel
Northern Ireland Building Regulations Advisory Committee
Northern Ireland Fishery Harbour Authority
Northern Ireland Housing Executive
Northern Ireland Tourist Board
Northern Ireland Transport Holding Company
Northern Ireland Water Council
Nuclear Electric PLC
Nuclear Powered Warships Safety Committee
Office of Population, Censuses and Surveys
OFFER
OFGAS
Offshore Industry Advisory Board
Offshore Safety and Technology Board
OFWAT
Oil and Pipelines Agency
Ordnance Survey
Overseas Development Administration
Parish Councils
Passenger Transport Authorities
Passenger Transport Executives
Peak Park Joint Planning Board
Poisons Board
Poisons Board (Northern Ireland)
Port Health Boards
Priorities Board for Research and Development in Agriculture and Food
Public Health Laboratory Service
Radio Communications Agency
Radioactive Waste Management Advisory Committee
Red Deer Commission
Regional, District and Islands Councils (Scotland)
Regional Panels for the Farming Industry
Renewable Energy Advisory Committee
River Purification Boards
Royal Commission on Environmental Pollution
Royal Botanic Gardens, Edinburgh

Royal Botanic Gardens, Kew
Royal Parks
Rural Development Commission
Science and Engineering Research Council
Scottish Agricultural Consultative Panel
Scottish Crop Research Institute
Scottish Enterprise
Scottish Fisheries Protection Agency
Scottish Natural Heritage
Scottish Nuclear
Scottish Office
Scottish Seed Potato Development Council
Scottish Tourist Board
Scottish Transport Group
Sea Fisheries Committees
Secretary of State's (Electricity) Fisheries Committee
Sheffield Urban Development Corporation
Standing Advisory Committee on Industrial Property
Statutory Waste Disposal Authorities
Street Works Advisory Committee
Technical Committee on Detergents and the Environment
Teeside Urban Development Corporation
Telford Urban Development Corporation
Trafford Park Urban Development Corporation
Transport and Road Research Laboratory
Tyne and Wear Urban Development Corporation
UK Atomic Energy Authority
Vehicle Inspectorate Executive Agency
Wales Tourist Board
Warren Spring Laboratory
Welsh Development Agency
Welsh Office

Appendix 3

USEFUL ADDRESSES

British Geological Survey
Central Enquiries Officer, Information Services Group
British Geological Survey, Keyworth, Nottingham NG12 5GG
Tel: 0115 9363 100

Compass (*Planning Magazine's* database)
Tel: 012452 310566

Countryside Commission
John Dower House, Crescent Place, Cheltenham,
Gloucestershire GL50 3RA
Tel: 01242 521381

Countryside Council for Wales
Plas Penrhos, Penrhos Road, Bangor, Gwynedd LL57 2LQ
Tel: 01248 370444

Department of the Environment
2 Marsham Street, London SW1P 3EB
Tel: 0171 276 3000

Department of the National Heritage
Heritage Division, 2–4 Cockspur Street, London SW1Y 5DH
Tel: 0171 211 6000

Eco-Management & Audit
Cameron Clark, The UK Competent Body for Eco-Management & Audit Scheme, Department of the Environment, Room C11/21, 2 Marsham Street, London SW1P 3EB
Tel: 0171 276 3377

English Heritage
Fortress House, 23 Savile Row, London W1X 1AB
Tel: 0171 973 3000

English Nature
Northminster House, Peterborough, Cambridgeshire PE1 1UA
Tel: 01733 340345

European Commission UK offices
Jean Monnet House, 8 Storey's Gate, London SW1P 3AT
Tel: 0171 973 1992

4 Cathedral Road, Cardiff CF1 9SG
Tel: 01222 371631

9 Alva Street, Edinburgh EH2 4PH
Tel: 0131 225 2058

European Parliament UK Information Office
2 Queen Anne's Gate, London SW1H 9AA
Tel: 0171 222 0411

Forestry Commission
231 Corstophine Road, Edinburgh EH12 7AT
Tel: 0131 334 0303

Goad Fire Insurance shops
Chas. E Goad Ltd., 8–12 Salisbury Square, Old Hatfield
Hertfordshire AL9 5BJ
Tel: 01707 271171

Health and Safety Executive
Baynards House, 1 Chepstow Place, Westbourne Grove
London W2 4TF
Tel: 0171 229 3456

Her Majesty's Inspectorate of Pollution
Romney House, 43 Marsham Street, London SW1P 3PY
Tel: 0171 276 0900

Inter-departmental Committee on the Redevelopment of Contaminated Land
Department of the Environment, Room A228, Romney House
43 Marsham Street, London SW1P 3PY
Tel: 0171 276 0900

Kelly's Directories
Editorial Department, Kelly's Directories (Reed Information Services),
Windsor Court, East Grinstead House, East Grinstead, West Sussex
RH19 1XB
Tel: 01342 326972

Land Use Survey
Land Use Research Unit, Department of Geography, King's College London,
Strand, London WC2R 2LS
Tel: 0171 873 2610

Ministry of Agriculture Fisheries and Food
Whitehall Place, London SW1A 2HH
Tel: 0171 270 8080

National Rivers Authority
Head Office (Bristol), Rivers House, Waterside Drive, Aztec West, Almondsbury, Bristol, BS12 4UD
Tel: 0117 9624400

Head Office (London), Eastbury House, 30–34 Albert Embankment, London SE1 7TL
Tel: 0171 820 0101

Ordnance Survey
Customer Orders, Romsey Road, Maybush, Southampton SO9 4DH
Tel: 01703 792338

Planning Inspectorate
Tollgate House, Houlton Street, Bristol BS2 9DJ
Tel: 0117 9878000

Soil Survey and Land Research Centre
The Archivist, Soil Survey and Land Research Centre, Silsoe Campus, Silsoe, Bedfordshire MK45 4DT
Tel: 01525 863000

United Kingdom Environmental Law Association
Secretary – Stephen Lloyd, Bates Wells & Braithwaite
61 Charterhouse Street, London EC1M 6HA
Tel: 0171 251 1122

Welsh Office/Swyddfa Gymreig
Cathays Park, Cardiff CF1 3NQ
Tel: 01222 82511

INDEX

References are to paragraph numbers.